5G and Beyond Wireless Transport Technologies

Douglas H Morais

5G and Beyond Wireless Transport Technologies

Enabling Backhaul, Midhaul, and Fronthaul

 Springer

Douglas H Morais
San Mateo, CA, USA

ISBN 978-3-030-74082-5   ISBN 978-3-030-74080-1   (eBook)
https://doi.org/10.1007/978-3-030-74080-1

This Springer imprint is published by the registered company Springer Nature Switzerland AG
The registered company address is: Gewerbestrasse 11, 6330 Cham, Switzerland

Preface

From the deployment of the very first mobile systems, there has been a need for connections to transport information between base stations and core operations. Such connections are referred to as backhaul. With 5G mobile networks, there currently exists the option to partition the base station into three physically separated subunits. As a result, with this partitioning, two new transport connections are required to interconnect these subunits. These new connections are referred to as midhaul and fronthaul. Collectively, we refer to backhaul, midhaul, and fronthaul connections as mobile transport connections. Such connections can be realized either optically or wirelessly. The majority will be backhaul, and it is estimated that by 2025 about 40% of backhaul connections globally will be wireless, which corresponds to about 60% when the fiber-dense countries of China, Japan, and South Korea are excluded. Thus, the need for, and the interest in, 5G wireless transport is high.

This text addresses primarily the key technologies that make today's 5G wireless transport connections possible and some that may find application in the future. Among these many technologies are high-order digital modulation, channel coding, phase noise suppression, co-channel dual polarization (CCDP) transmission accompanied by cross polarization interference cancellation (XPIC), line-of-sight MIMO, and band and carrier aggregation (BCA).

Several texts already exist that cover, individually, some of the key technologies addressed in this book, in addition to a multitude of others. No single one, however, as best the author can ascertain, addresses all the technologies covered here, all of which the author views to be key. Further, many, in general, cover technologies at a level that presupposes that the reader is already familiar with those technologies and seeks a deeper understanding. This book, on the other hand, presupposes only a general technical background in telecommunications and possibly no knowledge of some or all of the specific technologies covered.

The material presented is directed to industry professionals as well as academics. On the industry side, it should prove valuable to engineering managers, system engineers, technicians, and anyone who would benefit from a rounded understanding of the key technologies employed in 5G transport connections in order to more effectively execute their job. On the academic side, it should provide the upper undergraduate or graduate university student with a useful introduction to 5G transport connections. The material presented is intentionally not overly rigorous so as to be friendly to a wide audience and in keeping with a desire to convey a somewhat high-level view of the presented technologies. Mathematics, though clearly necessary for any meaningful study of the subject at hand, has been minimized. However, mathematical formulae applied in certain derivations in the text are provided in Appendix A. For those desiring to explore some or all of the material presented here in greater detail, several references are provided. A goal of this text is to allow the reader, if he/she so desires, to address the referenced and other more advanced material with confidence.

The author takes this opportunity to thank Jonas Hansryd of Ericsson for his invaluable help, advice, and support throughout the preparation of this book, and in particular for his assistance with Chaps. 1 and 7.

Finally, I wish to thank my editor at Springer, Mary James, for providing me with the opportunity to have this book published and for her instant attention to any question or issue that I presented to her throughout the preparation of this text.

San Mateo, CA, USA

Douglas H. Morais

Abbreviations and Acronyms

5G	Fifth Generation
3GPP	Third Generation Partnership Project
5GC	5G Core
ADC	Analog to Digital Converter
AIC	Air Interface Capacity
AMC	Adaptive Modulation and Coding
ARQ	Automatic Repeat reQuest
ATE	Adaptive Transversal Equalizer
BB	Baseband
BBP	Baseband Processor
BCA	Band and Carrier Aggregation
BEC	Binary Erasure Channel
BER	Bit Error Rate
BPSK	Binary Phase Shift Keying
CCDP	Co-channel Dual Polarization
CMA	Constant-Modulus Algorithm
CN	Check Node
CPRI	Common Public Radio Interface
C-RAN	Centralized RAN
CRC	Cyclic Redundancy Check
CU	Central Unit
DAC	Digital to Analog Converter
DFE	Decision Feedback Equalizer
DL	Downlink
DSBSC	Double-Sideband Suppressed Carrier
DU	Distributed Unit
eCPRI	Enhanced CPRI
EIRP	Equivalent Isotropically Radiated Power
eMBB	Enhanced Mobile Broadband
FEC	Forward Error Correction

FIR	Finite Impulse Response
IAB	Integrated Access and Backhaul
IF	Intermediate Frequency
IP	Internet Protocol
ISI	Inter Symbol Interference
ITU	International Telecommunications Union
LDPC	Low Density Parity Check
LLR	Log Likelihood Ratio
LMS	Least Mean Square
LoS	Line-of-Sight
MAC	Medium Access Control
MCMA	Modified Constant-Modulus Algorithm
MIMO	Multiple-Input Multiple-Output
ML	Maximum Likelihood
mMTC	Massive Machine Type Communications
MPLS	Multi-protocol Label Switching
MTU	Maximum Transfer Unit
OAM	Orbital Angular Momentum
OMT	Orthomode Transducer
PAM	Pulse Amplitude Modulation
PCM	Parity Check Matrix
PDCP	Packet Data Convergence Protocol
PHY	Physical
PN	Phase Noise
PSD	Power Spectral Density
QAM	Quadrature Amplitude Modulation
QoS	Quality of Service
QPSK	Quadrature Phase Shift Keying
RAN	Radio Access Network
RF	Radio Frequency
RLC	Radio Link Control
RLS	Recursive Least Square
RRC	Root Raised Cosine
RS	Reed Solomon
RTP	Real-Time Transport Protocol
RU	Radio Unit
SCMA	Simplified Constant-Modulus Algorithm
SDAP	Service Data Application Protocol
SNR	Signal-to-Noise Ratio
STE	Space Time Equalizer
TCP	Transmission Control Protocol
TVE	Transversal Equalizer
UDP	User Datagram Protocol
UL	Uplink

UMTS	Universal Mobile Telecommunications System
URLLC	Ultra-Reliable and Low Latency Communications
VCO	Voltage Controlled Oscillator
VN	Variable Node
voIP	Voice over IP
XPD	Cross Polarization Discrimination
XPIC	Cross-Polarization Interference Cancellation
ZF	Zero-Forcing

Contents

1 5G Architecture and the Roll of Wireless Transport 1
  1.1 Introduction 1
  1.2 5G Usage Scenarios and Top-Level Requirements 1
  1.3 5G Network Overview 2
  1.4 5G Transport Network Components (Backhaul, Midhaul,
      Fronthaul) 4
  1.5 Transport Realization Options 7
      1.5.1 Fiber 7
      1.5.2 Free Space Optics 8
      1.5.3 Wireless 8
  1.6 Key Wireless Transport Technologies in Support
      of 5G Networks 16
  1.7 Summary 18
  References 18

2 5G Transport Payload: Ethernet-Based Packet-Switched Data 19
  2.1 Introduction 19
  2.2 TCP/IP 20
      2.2.1 Application Layer Protocol 20
      2.2.2 Transport Layer Transmission Control Protocol 21
      2.2.3 Transport Layer User Datagram Protocol 21
      2.2.4 Internet Layer Protocol 22
      2.2.5 Data Link Layer Ethernet Protocol 24
      2.2.6 Multi-Protocol Label Switching (MPLS) 27
  2.3 Voice over IP (VoIP) 27
  2.4 Video over IP 28
  2.5 Header Compression 29
  2.6 Payload Compression 30
  2.7 Summary 31
  References 31

3   The Fixed Wireless Path   33
    3.1   Introduction   33
    3.2   Antennas   34
            3.2.1   Introduction   34
            3.2.2   Antenna Characteristics   34
            3.2.3   Typical Point-to-Point Wireless Antennas   36
    3.3   Free Space Propagation   40
    3.4   Line-of-Sight Non-Faded Received Signal Level   42
    3.5   Fading Phenomena   44
            3.5.1   Atmospheric Effects   44
            3.5.2   Terrain Effects   52
            3.5.3   Signal Strength Versus Frequency Effects   59
            3.5.4   Cross-Polarization Discrimination Degradation
                    due to Fading   66
    3.6   External Interference   67
    3.7   Outage and Unavailability   68
    3.8   Diversity Techniques for Improved Reliability   69
            3.8.1   Space Diversity   70
            3.8.2   Angle Diversity   72
    3.9   Summary   73
    References   73

4   Digital Modulation: The Basic Principles   75
    4.1   Introduction   75
    4.2   Baseband Data Transmission   75
    4.3   Linear Modulation Systems   82
            4.3.1   Double-Sideband Suppressed Carrier (DSBSC)
                    Modulation   82
            4.3.2   Binary Phase-Shift Keying (BPSK)   85
            4.3.3   Quadrature Amplitude Modulation (QAM)   88
            4.3.4   Quadrature Phase-Shift Keying (QPSK)   89
            4.3.5   High-Order 2^{2n}-QAM   93
            4.3.6   High-Order 2^{2n+1}-QAM   96
            4.3.7   Peak-to-Average Power Ratio   100
    4.4   Transmission IF and RF Components   102
            4.4.1   Transmitter Upconverter and Receiver Downconverter . . .   103
            4.4.2   Transmitter RF Power Amplifier and Output
                    Bandpass Filter   105
            4.4.3   The Receiver "Front End"   105
    4.5   Modem Realization Techniques   107
            4.5.1   Scrambling/Descrambling   107
            4.5.2   Carrier Recovery   109
            4.5.3   Timing Recovery   113
    4.6   Summary   114
    References   115

5 Performance Optimization Techniques 117
   5.1   Introduction 117
   5.2   Forward Error Correction Coding 118
          5.2.1   Introduction 118
          5.2.2   Block Codes 119
          5.2.3   Classical Parity-Check Block Codes 123
          5.2.4   Low-Density Parity-Check (LDPC) Codes 124
          5.2.5   Reed-Solomon (RS) Codes 127
          5.2.6   LDPC and RS Codes in Wireless Transport 133
          5.2.7   Polar Codes 133
   5.3   Block Interleaving 138
   5.4   Puncturing 140
   5.5   Adaptive Modulation and Coding (AMC) 141
   5.6   Power Amplifier Linearization Via Predistortion 142
   5.7   Phase Noise Suppression 143
   5.8   Quadrature Modulation/Demodulation Imperfections
         Mitigation 148
          5.8.1   Transmitter Quadrature Error Mitigation 149
          5.8.2   Transmitter I/Q Balance Error Mitigation 150
          5.8.3   Transmitter Residual Error Mitigation 151
          5.8.4   Receiver Quadrature Imperfections Mitigation 151
   5.9   Adaptive Equalization 153
          5.9.1   Introduction 153
          5.9.2   Time-Domain Equalization 153
   5.10 Summary 164
   References 164

6 Non-Modulation-Based Capacity Improvement Techniques 167
   6.1   Introduction 167
   6.2   Co-Channel Dual Polarization (CCDP) Transmission 167
   6.3   Line-of-Sight Multiple-input Multiple-output (LoS MIMO) 170
          6.3.1   Introduction 170
          6.3.2   LoS MIMO Fundamentals 171
          6.3.3   Optimal Antenna Separation 173
          6.3.4   Non-optimal Antenna Separation 175
          6.3.5   LoS MIMO Equalization 176
          6.3.6   Increasing Channel Capacity Via the Simultaneous
                  Use of CCDP/XPIC and LoS MIMO 177
   6.4   Orbital Angular Momentum Multiplexing 179
          6.4.1   Introduction 179
          6.4.2   OAM Structure and Characteristics 181
          6.4.3   OAM Mode Generation and Multiplexing/
                  Demultiplexing 182
   6.5   Band and Carrier Aggregation 183
   6.6   Summary 186
   References 187

**7 Transceiver Architecture, Link Capacity, and Example
Specifications** 189
   7.1   Introduction 189
   7.2   Basic Transceiver Architecture and Structural Options 189
       7.2.1   The Baseband Processor 191
       7.2.2   The IF Processor 193
       7.2.3   The Direct Conversion RF Front End 194
       7.2.4   The Heterodyne RF Front End 194
       7.2.5   Antenna Coupling 195
       7.2.6   The Antenna 198
   7.3   Link Capacity Capability 198
   7.4   Example Specifications and Typical Path Performance
of an 80 GHz (E-Band) Link 201
   7.5   Example Specifications and Typical Path Performance
of a 32 GHz Link 204
   7.6   Conclusion 207
   References 208

Appendices 209

Appendix A 209

Appendix B 211

Appendix C 216

Appendix D 222

Appendix E 226

Index 229

About the Author

Douglas H. Morais has decades of experience in the wireless communications field that encompasses product design, engineering management, executive management, consulting, and short course lecturing. He holds a B.Sc. from the University of Edinburgh, Scotland; an M.Sc. from the University of California, Berkeley; and a Ph.D. from the University of Ottawa, Canada, all in electrical engineering. Additionally, he is a graduate of the AEA/Stanford Executive Institute, Stanford University, California, is a Life Senior member of the IEEE, and a member of the IEEE Communications Society.

Dr. Morais has authored several papers on wireless digital communications; holds three US patents, one on point-to-multipoint wireless communications and two on digital modulation; and has authored two books: *Fixed Broadband Wireless Communications* published by Prentice Hall PTR and *Key 5G Physical Layer Technologies* published by Springer.

Chapter 1
5G Architecture and the Roll of Wireless Transport

1.1 Introduction

This text deals primarily with the key technologies that support *wireless transport* (Sect. 1.4) in *fifth generation* (5G) and beyond mobile systems. Before we address these technologies, however, many readers will find it helpful if certain ancillary information is presented. To this end, in this chapter, we first review the stated 5G usage scenarios and 5G top-level requirements. Next, a very high-level description of the overall 5G architecture is provided and the location and role of the transport network in this architecture indicated. This is followed by an outline of the various transport connection realization options, including fiber, wireless, and free space optics. The wireless transport outline includes a short recital of its evolution, a description of the applicable frequency bands, and an introduction to Integrated Access and Backhaul. Finally, a listing and short description of the key technologies that support wireless transport and that are addressed in this text is given.

1.2 5G Usage Scenarios and Top-Level Requirements

The ITU in 2015 defined fifth generation (5G) systems as those that meet its IMT-2020 requirements. IMT-2020 envisaged the support of many usage scenarios, three of which it identified: enhanced mobile broadband (eMBB), ultra-reliable and low latency communications (URLLC), and massive machine-type communications (mMTC):

– eMBB is the natural evolution of broadband services provided by 4G networks. It addresses human-centric use cases and applications, enabling the higher data rates and data volumes required to support ever-increasing multimedia services.

D. H. Morais, *5G and Beyond Wireless Transport Technologies*,
https://doi.org/10.1007/978-3-030-74080-1_1

- URLLC addresses services requiring very low latency and very high reliability. Examples are factory automation, self-driving automobiles, and remote medical surgery.
- mMTC is purely machine-centric, addressing services that provide connectivity to a massive number of devices, driven by the growth of the Internet of Things (IoT). Such devices are expected to communicate only sporadically and then only a small amount of data. Thus, support of high data rates here is of less importance.

Among the many IMT-2020 envisaged requirements are (a) the capability of providing a peak *download* (DL) data rate of 20 Gb/s and a peak *upload* (UL) data rate of 10 Gb/s, (b) user experienced DL data rates of up to 100 Mb/s and UL data rates of up to 50 Mb/s, (c) over-the-air latency of 1 ms, and (d) operation during mobility of up to 500 km/hr.

3GPP commenced standardization work on its 5G wireless access technology in 2016 and labeled it *New Radio* (NR). It decided to address 5G in two phases. The first phase, Release 15 (Rel-15), only addressed eMBB and URLCC and was released in 2019. Rel-16 addresses all IMT-2020 scenarios, meets all key capabilities, and was released in 2020. The first version of Rel-15, ver. 15.1.0, allowed operation in two broad frequency ranges: sub-6 GHz (450 MHz to 6 GHz), referred to as FR1, and millimeter wave (24.25 GHz to 52.6 GHz), referred to as FR2. Rel-15, ver. 15.5.0 expanded FR1's range (410 MHz to 7.125 GHz) while leaving the FR2 range unchanged. Networks with performance approaching that of 5G NR Rel-15 specifications began operating in 2019.

From a wireless transport perspective, the most pertinent of the 5G specifications is the extremely high data rates envisaged for some of the usage scenarios. This is because these rates directly affect the transport rates required. Fiber optic communication can handle these higher rates in its stride. For wireless transport, however, channels with greater bandwidth and/or the application of advanced technologies are called for.

1.3   5G Network Overview

The overall 5G NR network, specified by the Third Generation Partnership Project (3GPP), comprises the core network (5GC) and the *radio access network* (RAN). The core network is responsible for those functions not related to radio access but required to provide a complete network. The RAN is responsible for all radio-related functions such as mapping user data streams according to their defined quality of service (QoS), IP header compression, scheduling, coding, modulation, physical transmission to and from the end users, etc. A detailed description of the RAN can be found in [1]. Its user plane protocol layer structure is shown in Fig. 1.1 where SDAP stands for *Service Data Application Protocol*, PDCP stands for *Packet Data Convergence Protocol*, RLC stands for *Radio Link Control*, MAC stands for *Medium Access Control*, and PHY stands for *Physical*.

Fig. 1.1 NR user plane protocol layer structure

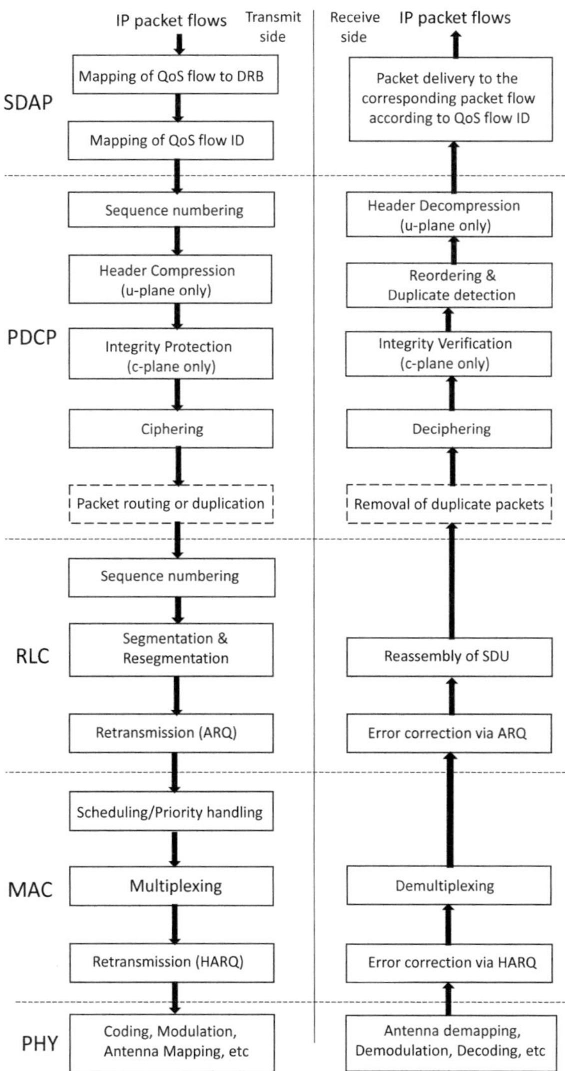

For our purposes, we note that the 3GPP defined for 5G a RAN architecture where the base station (gNB) functionality is split into two logical units. By logical we mean that a split between functionality is defined but does not necessarily imply a physical split. The two units are the *central unit* (CU) and the *distributed unit* (DU). The CU is connected to the 5GC via the NG interface. The 3GPP studied eight functional splits between the CU and the DU in [2]. It selected Option 2 functional split (PDCP/RLC) for the interface between the CU and DU, this interface being labeled the F1 interface. Within the CU and DU are different layers of functionality. In the DU the lowest layer of functionality is the physical layer which handles among

Fig. 1.2 Simplified 5G Network architecture

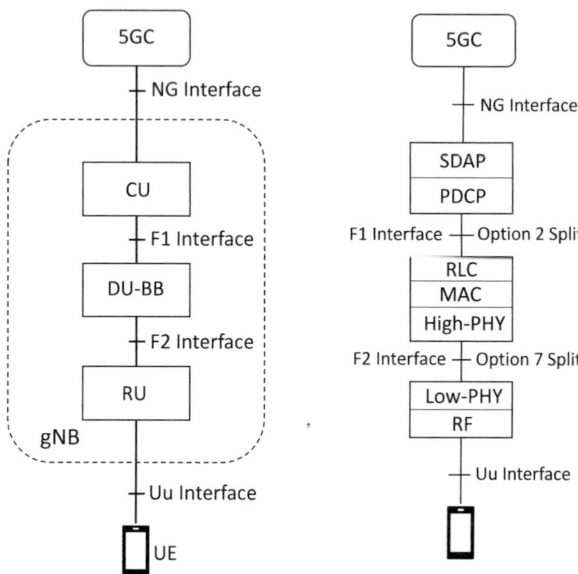

other functions the conversion of data from baseband (BB) format to RF and vice versa. Often the RF functionality is physically separated from the BB functionality. Is convenient therefore to consider the DU as two logical units, namely the DU-BB unit and the *radio unit* (RU) connected via what we shall term the F2 interface, this interface not currently being an officially specified one by the 3GPP. The RU implements the RF functions and may or may not contain some physical layer functions. It interfaces with the user equipment (UE) via the Uu interface. Figure 1.2 shows a highly simplified 3GPP 5G architecture. On the left, we see the CU, DU-BB, and RU divisions. On the right, we see these divisions relative to the 5G NR user plane protocol stack for a specific case where the RU contains some physical layer functions referred to as "Low-PHY" and the DU-BB contains the remaining physical layer functions referred to as "High-PHY." This split corresponds to Option 7 functional split as defined by the 3GPP in [2].

1.4   5G Transport Network Components (Backhaul, Midhaul, Fronthaul)

Shown in Fig. 1.3 are several physical arrangements via which the 5GC can be connected with the UE. The transport network between the 5GC and the gNB's CU which implements the NG interface is referred to as *backhaul*. Backhaul payload is IP packet-based and so transportable over Ethernet-based (Sect. 2.2) transport links. We note that all physical arrangements shown in Fig. 1.3 employ backhaul. In Fig. 1.3b, however, there is a physical functional split between the CD and the

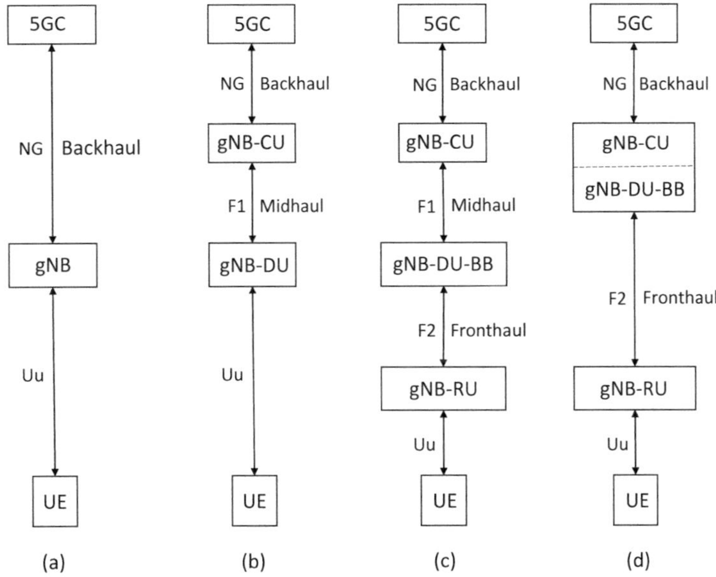

Fig. 1.3 Several physical arrangements for 5GC to UE connection

DU. This arrangement thus calls for a transport network between the CU and the DU which implements the F1 interface, and this network is referred to as *midhaul*. As with backhaul, midhaul payload is packet-based and so can be transported over packet-based transport mechanisms such as IP/Ethernet. In Fig. 1.3c, there is a physical functional split between the DU-BB and the RU thus calling for a transport network to implement the F2 interface between these units. This network is referred to as *fronthaul*. Finally, in Fig. 1.3d we see an arrangement where only backhaul and fronthaul are utilized. Backhaul, midhaul, and fronthaul are here referred to collectively as X-haul [3].

We note that the structure shown in Fig. 1.3a is referred to as a *distributed RAN* (D-RAN) architecture as the data flow to and from the core network is distributed to each physically undivided gNB. The structures shown in Fig. 1.3b, c, and d, on the other hand, are referred to as *centralized RAN* (C-RAN) architectures. Here many of the gNB functions are centralized in the CU which is placed in a more central location enabling more optimum network coordination and where one CU can handle a large number of DUs and RF heads. Figure 1.4 shows a typical transport network topology where fronthaul and midhaul are applied. In the direction from RU to the 5GC, data streams from several RUs are transported via fronthaul to a DU-BB where they are aggregated and transported via midhaul to a CU where midhaul received data is aggregated and transported via backhaul to the 5GC. The reverse takes place in the 5GC to RU direction.

Fronthaul connections were first implemented in third generation (3G) mobile networks. There, the radio unit was referred to as the remote radio head (RRH) and contained no physical functions, only RF-related ones such as baseband to RF and

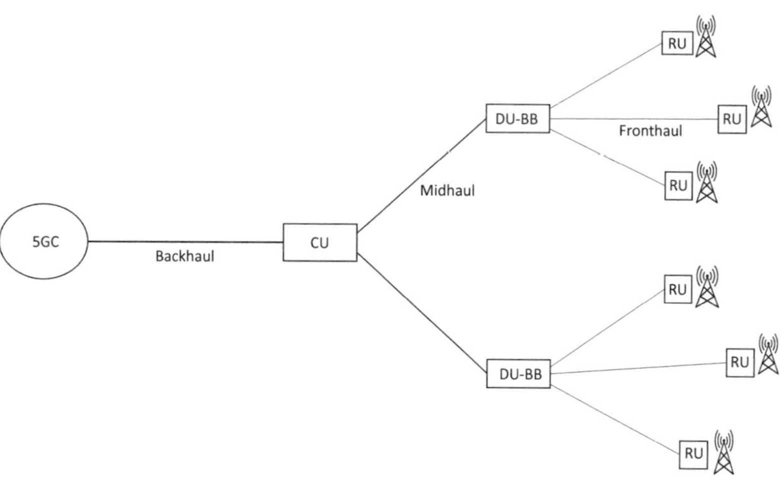

Fig. 1.4 Typical Transport Network Topology

RF to baseband conversion and RF amplification. The connection between it and the rest of the base station, referred to as the baseband unit (BBU), was via Common Public Radio Interface (CPRI). This interface, in effect at the Option 8 functional split as per [2], was designed to transport in a continuous fashion digitized time-domain samples of the baseband signal between the RRH and BBU. CPRI supports several constant bit rate options, ranging from a low of 614.4 Mb/s to a current high of 24.3302 Gb/s. At first glance, the highest rate would seem to be adequate for most situations. However, CPRI-based fronthaul traffic load is proportional to the number of antennas and the RAN RF bandwidth, independent of what portion of that bandwidth is currently being used. 5G RAN often employs, via massive MIMO [1], a high number of antennas and occupies large RF bandwidth. For example, a system with 32 antennas per sector and a bandwidth of 100 MHz would require a CPRI fronthaul rate of approximately 300 Gb/s per sector and thus would not be supported by CPRI.

As a result of CPRI's inability to support most 5G requirements, the industry partners who developed and maintain the CPRI specification introduced enhanced CPRI (eCPRI) [4] in 2017. With eCPRI, the functional split between the RU and the DU is Option 7 as per [2] where certain physical layer functions are moved from the DU-BB to the RU, these functions being referred to as low-PHY, the rest of the physical functions remaining in the DU being referred to as high-PHY. The changes made result in a large amount of the overhead and non-current user-dependent data being eliminated and thus a significant reduction of the required data rate compared to CPRI. Further, it lowers the latency and jitter for high priority traffic, and, very importantly, its traffic can be transported on an Ethernet (Sect. 2.2.5) based link along with other traffic in the same network, including other eCPRI data. It removes the constant bit rate requirement required with CPRI as the rate now scales with the currently used antenna bandwidth which, in the 5G NR multicarrier transmission

Table 1.1 Latency parameters of X-Haul connections

	Backhaul	Midhaul	Fronthaul
Latency	< 10 ms	< 5 ms	< 100 µs

system, in turn, scales with the current user traffic. A second option for fronthaul interface is the Radio over Ethernet (RoE) interface, developed by the IEEE 1914.3 Working Group [5]. Though capable, as the name implies, of Ethernet transport, the main drawback of RoE is that it affords minimal bandwidth reduction compared to CPRI. Thus, eCPRI is the most efficient Ethernet-based 5G fronthaul option. Utilizing eCPRI and aggregating packets from the various antennas at the antenna site is the most efficient approach, reducing capacity demands compared to CPRI or RoE by 60–80% depending on the radio configuration [6].

For the same information transmission on fronthaul, midhaul, and backhaul, the backhaul and midhaul data rates are likely to be very similar, but the fronthaul data rate, despite the improvement afforded by eCPRI, is likely to be about ten times that at midhaul and backhaul. In a dense urban environment, fronthaul transport capacity could thus be in the high tens of Gb/s to the low hundreds of Gb/s. Further, as shown in Table 1.1, fronthaul requires very low latency (< about 100 µs) compared to midhaul (<about 5 ms) and backhaul (< about10 ms).

1.5 Transport Realization Options

The high data rates supported by 5G NR coupled with the requirement for low latency place much more challenging requirements on 5G NR transport networks compared to those called for by 4G networks when the need is to provide the fully defined features of 5G. For 5G NR deployments where there is a high density of small cells and very high traffic data, transport networks will likely over time need to support up to in the order of a hundred Gb/s of data. The technology options likely for consideration to meet 5G NR transport requirements are fiber optic cable, wireless, and, possibly in the future, free space optics. Following is a brief description of these options.

1.5.1 Fiber

Of all the transport options, fiber can provide the highest capacity, lowest bit error rate, low latency, and the furthest distance before signal retransmission. An existing link can keep pace with growing requirements through multiplexing different wavelengths on existing strands or by pulling new strands through existing conduit. Further, no licensing of the spectrum is required. Fiber is thus the first choice for transport connections, if available. Unfortunately, it is not always available. And should it not be available, new deployment typically takes several months and comes at a high cost which includes not only the cost of the cable and supporting electronic equipment but also the cost of right-of-way acquisition, trenching, splicing, etc. Thus, if cable is not already available, other options receive serious consideration. Fiber today connects in the region of 40% of all backhaul networks, excluding the fiber dense countries China, Japan, and South Korea.

1.5.2   Free Space Optics

Free space optics (FSO) is somewhat similar to optical fiber communications but, on the other hand, is "wireless." As with fiber, the transmission signal is generated by a laser but transmitted over free space in the infrared or visible range, the transmitted laser beams being very narrow. FSO affords a transmission bandwidth in the range of hundreds of GHz and hence has the potential to provide very high data rates, from the tens of Gb/s to in excess of 100 Gb/s. FSO, nonetheless, has limitations. It requires a clear line-of-sight not subject to temporary blockage. It calls for precise alignment of the laser transmitter and optical receiver and thus minor movement can result in significant signal attenuation. It suffers from fading due to dense fog as the size of fog particles is comparable to the optical wavelengths. Signal attenuation due to fog can be in excess of 100 dB/km. Reliable communication is thus typically limited to a few hundred meters. As a result of these limitations, FSO, though promising, is not really a player at present in the transport network market. Over time, however, this could change given the increasing need in highly dense urban environments for very high data rate capability over very short distances and FSO's ability to meet this need.

1.5.3   Wireless

Wireless is and will continue to be a meaningful player in the mobile network backhaul market, being estimated to support, by 2025, 38% of backhaul connections globally, this corresponding to 62% when excluding the fiber dense countries China, Japan, and South Korea [7], and may over time play a similar role in the newly emerging 5G midhaul market. This participation will be aided by operation at millimeter wave frequencies with very wide bandwidth and hence high data rate capability. Wireless will be used mainly for non-aggregated backhaul and midhaul links close to Radio Units in urban and dense urban areas and aggregated backhaul and midhaul links in suburban and rural areas. As mentioned above, fronthaul capacity is likely to be in the tens of Gb/s and thus best served by fiber. However, when the time to market is important and cable is not readily available, high-capacity millimeter-wave links may play a role in the fronthaul market.

A wireless network can be structured as point-to-point links, point-to-multipoint links, or as a mesh network. However, the point-to-point structure is by far the most popular. Its over-the-air latency (3.3 µs/km) is less than 70% that of fiber as air is a faster medium than glass. Further, it traverses over the shortest distance between the point of transmission and reception. However, each terminal and each repeater adds to the total latency and fiber can travel much greater distances than wireless without needing a repeater, thus if a wireless backhaul section has several repeaters it may have a higher total latency than fiber. Its major attractions are ease and short time of deployment, cost-effectiveness relative to fiber, and allowing the operator to own and control the backhaul network rather than leasing service, which is typically the case with fiber. That said, there are numerous factors that must be taken into account if wireless is to be successfully deployed. Among these are spectrum availability, propagation environment, interference conditions, line-of-sight availability, etc.

Wireless backhaul can be accomplished via networks that operate independently of the mobile access network or via ones where the access network and some parts of the backhaul network are integrated as discussed in Sect. 1.5.3.4 below. For analysis purposes, it is convenient to view independent wireless backhaul, midhaul, and fronthaul as falling into one of two classifications, i.e., either one utilizing traditional bands or one utilizing nontraditional millimeter-wave bands. Traditional bands occupy from 6 GHz to 42 GHz. The nontraditional millimeter-wave bands cover from 57 GHz to 175 GHz and because of the high susceptibility of wireless systems operating in these bands to atmospheric conditions, such systems can rightly be viewed as belonging to a separate class. In the following sections, we will look more closely at the traditional bands that are most suitable for backhaul and midhaul and at the nontraditional millimeter-wave bands that are suitable for backhaul, midhaul, and fronthaul. We will then look at backhaul where access and some parts of backhaul are integrated. However, before we do that, we will briefly review the evolution of wireless transport for its earliest implementation to the present time.

1.5.3.1 Wireless Transport Evolution

Wireless mobile transport dates from the beginning of cellular mobile systems in the early 1980s and commenced as solely backhaul. The *first generation* (1G) systems were analog but used point-to-point digital microwave for the backhaul links connected to or close to the base stations. These links transported *time division multiplexed* (TDM) data and had capacities in the multiple T1/E1 (1.544/2.048 Mb/s) range and operated initially in or close to the 2 GHz band but migrated slowly to the 18 to 23 GHz range. As links from the base stations were aggregated as they approached the Main Telephone Switching Office (MTSO) the required data transmission rate increased. As a result, the backhaul section close to the MTSO required rates of multiple DS3/E3 (44.736/34.368 Mb/s). This high-capacity transport was provided either by microwave links operating in the 6–11 GHz frequency range or via fiber, typically provided by the local telephone company.

Second generation (2G) systems appeared in the early 1990s. Though, like 1G systems, they were primarily aimed at voice services, they utilized digital modulation, allowing a higher voice capacity and support of low-rate data applications. As an example of data capabilities, the original *Global System for Mobile Communications* (GSM) standard supported circuit-switched data at 9.6 kb/s. By the late 1990s, however, with the introduction of *Enhanced Data Rate for GSM Evolution* (EDGE), user rates of between 80 and 120 kb/s were supported. Like 1G systems, these systems used microwave radio for backhaul links connected to or close to the base stations, these links having capacities in the multiple T1/E1 range and with operating frequencies ranging from 18 to 38 GHz. Higher data rate backhaul connection was provided predominantly by fiber.

Third generation (3G) systems became available in the early 2000s and represented a significant leap over 2G ones. These systems were conceived to deliver a wide range of services, including telephony, higher speed data than available with 2G, video, paging, and messaging. All the 3G systems utilized *Code Division Multiple Access* (CDMA) technology, provided significant increase in voice capacity, and offered much higher data rates over both circuit-switched and packet-switched bearers. These systems initially all provided peak *downlink* (DL) and *uplink* (UL) rates in the high kb/s to the very low Mb/s rates. These rates increased over time, but the breakout technology was *Universal Mobile Telecommunications System* (UMTS) whose continued evolution led to extremely high data rates. Its first release, Rel. 99, was labeled WCDMA, and offered downlink (DL) and Uplink (UL) data rates of 0.384 Mb/s and specified Asynchronous Transport Mode (ATM) [8], a packet data type, as its transport technology. The physical layer used to effect transport, however, was standard TDM based T1/E1. Thus, T1/E1 microwave links were used for backhaul close to the base station. The following CDMA-based UMTS releases increased data rates to tens of Mb/s, and IP/Ethernet was specified as a transport alternative to ATM. TDM based platforms didn't easily scale to support the ever-increasing data rates. As a result, hybrid solutions were introduced where TDM type of traffic such as voice data was carried on TDM T1/E1 transport links, but the high-speed data was transported on dedicated Ethernet-based IP links. Wireless backhaul solutions thus also became hybrid, handling TDM and packet-based traffic on two different physical channels.

UMTS in 2009 released a system based on *orthogonal frequency division multiplexing* (OFDM) technology. This system was labeled *Long Term Evolution* (LTE). It was designed to permit a DL maximum data rate of 300 Mb/s and a UL maximum data rate of 75 Mb/s and commenced service in 2009. LTE wasn't true *fourth generation* (4G) as defined by the ITU, but such a system became available in the early 2010s as an evolution of LTE under the nomenclature *LTE-Advanced*. LTE and LTE-Advanced represented a key departure from its predecessors in that its connection to the core network was all IP, this connection accomplished via Ethernet links. Wireless backhaul used in these systems is thus comprised of all Ethernet links, e.g., Fast Ethernet (100 Mb/s) or Gigabit Ethernet (1Gb/s). UMTS also, for the first time, specified a logical separation of the base station into a baseband unit (BBU) and a remote radio head (RRH). Thus, the operator could choose to physically separate the BBU and the RRH via a fronthaul link. As indicated in Sect. 1.4 above, this link used CPRI.

As discussed in Sect. 1.2 above, fifth generation (5G) systems commenced service in 2019 with the goal of being able to provide a) a peak DL data rate of 20 Gb/s and a peak UL data rate of 10 Gb/s, b) user experienced DL data rates of up to 100 Mb/s and UL data rates of up to 50 Mb/s, c) and over-the-air latency of 1 ms. Like 4G, 5G backhaul networks are all IP/Ethernet-based, but now, the transport network may also comprise IP/Ethernet-based midhaul and fronthaul. Further, much higher data rates must be handled. Wireless Ethernet backhaul and midhaul links will need to handle capacities from about 200 Mb/s for rural relatively low data rate carrying networks up to about 20+ Gb/s in the future for dense urban high data rate carrying networks. Wireless Ethernet fronthaul links will need to transport the highest data rates of all and are likely to be used only in situations where fiber is not readily available.

1.5.3.2   Traditional Bands

Typical operating frequencies for independent transport in the microwave traditional bands are those close to 6, 7/8,11, 13, 15, 18, 23, 26, 28, 32, 38, 40, and 42 GHz. Figure 1.5 [9] shows the extent of microwave backhaul by band and region, the size of each circle being relative to the number of microwave hops in the region. The 26, 28, 38, and 42 GHz bands, however, are now designated in many countries for 5G mobile access. Further, at the World Radiocommunications Conference 2019 (WRC-19), the 26 GHz (24.25–27.5 GHz), and the 40 GHz (37–43.5 GHz) bands were identified for future use as 5G mobile access bands. Thus, the use in transport networks not integrated with the access network of the bands between 26 and

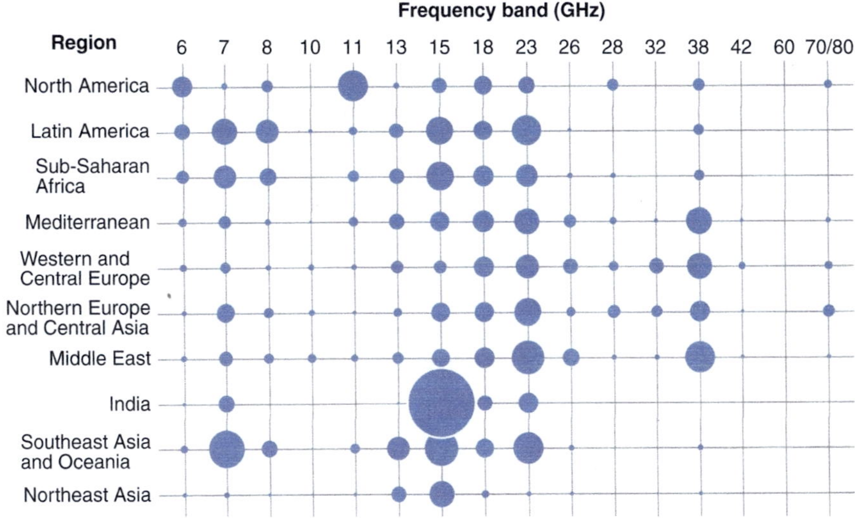

Fig. 1.5 Global use of microwave backhaul (with the permission of Ericsson)

42 GHz, with the exception of the 32 GHz band, could become limited over time as they more and more become applied to 5G access. The 32 GHz band, not being identified for 5G access, will thus remain active and become increasingly essential for backhaul.

The bands below about 12 GHz are indispensable for long-haul backhaul because of their better propagation characteristics relative to higher frequency bands. Among those available below 12 GHz bands, the 6 GHz band has been the workhorse one, being heavily used in many countries. However, recent actions have put the mobile transport application of this band under threat. The United States has decided to allow unlicensed use in the 5.925–7.125 GHz band for services such as Wi-Fi. The specifications on the unlicensed use are designed to minimize interference with licensed use. Specifically, two different types of unlicensed operation are authorized, indoor low-power and outdoor standard-power. The indoor units have a maximum allowed Equivalent Isotropic Radiated Power (EIRP) (Sect. 3.2.2) of 30 dBm (1 Watt) and the standard-power units are allowed a maximum EIRP of 36 dBm (4 Watts). Low-power operation is permitted all across the band, and its power level is such that it is unlikely to cause interference with licensed links. For those parts of the band that support a large number of high-reliability point-to-point links such as mobile backhaul links, standard power operation is permitted under the control of an automated frequency coordination (AFC) system. The AFC system determines the frequencies on which standard-power devices can operate without likely causing harmful interference to incumbent microwave receivers. It does this by referring to a database of incumbent licensed systems and then makes only available for use by standard-power devices those frequencies that won't interfere with incumbent links. However, this cannot fully guarantee that outages due to interference will not occur. In Europe also, the threat from unlicensed use is becoming an issue as consideration is being given to allow unlicensed operation in the 5.925–6.425 GHz band. The uncertainty is further compounded by the fact that the WRC at its next conference in 2023 will consider IMT (5G) identification for the 6.425–7.125 GHz band. The alternative band is the 7/8 GHz (7.125–8.5 GHz) band, but in the United States, unfortunately, this band is reserved for federal use. Outside of the United States, therefore, the 7/8 GHz will likely become increasingly essential for long-haul backhaul.

The bandwidth per assigned channel tends to increase with frequency. Thus, between 6 and 11 GHz the typical maximum operating bandwidth is 28 MHz, at 13 and 15 GHz it is 56 MHz, and at frequencies from 18 GHz to 42 GHz it is 112 MHz. As a result, relative to the sub 12 GHz bands, greater capacity tends to be available in the 13 and 15 GHz bands, and even greater still in 18 GHz and above frequency bands. The trade-off, however, is distance. Links operating between 6 and 11 GHz can typically operate at high reliability over much longer distances (up to about 30 km) than those in the region of 20 GHz (here up to about 10 km) and those in the region of 40 GHz (here up to about 5 km). Typically, single-channel links operating between 6 and 11 GHz can support data rates up to about 550 Mb/s whereas those operating between 18 and 42 GHz can support data rates up to about 4.4 Gb/s. Table 1.2 shows typical maximum data rates for the traditional bands,

Table 1.2 Some typical characteristics of traditional bands

Frequency band	Typical max. bandwidth	Cell location	CCDP[a]	2 × 2 MIMO[b]	Typical max. Data rate per RF channel
6–11 GHz	28 MHz	Rural	x	x	~1.1 Gb/s
13–15 GHz	56 MHz	Suburban/ semirural	x	x	~2.2 Gb/s
18–42 GHz	112 MHz	Suburban/ semirural	x	x	~4.4 Gb/s

[a]Co-channel dual polarization (CCDP) increases capacity by a factor of 2
[b]2 x 2 Multiple-Input Multiple-Output (MIMO) increases capacity by a factor of 2

typical associated maximum bandwidth, cell location, and the impact of co-channel dual polarization (CCDP), which is described in Sect. 6.2, and Multiple-Input Multiple-Output (MIMO), which is described in Sect. 6.3. Even higher rates are achievable via band and carrier aggregation (Sect. 6.5), i.e., transmitting over more than one channel in a band or in multiple bands. Data rates attainable in the traditional bands make them suitable for backhaul and midhaul operations. Fronthaul operation here is unlikely, however, given the very high data rates involved.

1.5.3.3   Nontraditional Millimeter-Wave Bands

Backhaul, independent of the access network, as well as midhaul and fronthaul, can operate in the nontraditional millimeter-wave bands that cover 57–175 GHz. As shown in Fig. 1.6 there are currently four bands in this classification. The first is referred to as the *V-Band* which extends from 57 to 71 GHz (57–66 GHz Europe, 57–71 GHz United States). It suffers large attenuation from oxygen and water in a narrow band around 60 GHz (see Fig. 3.16) limiting its use for transmission in this narrow-band to very short-range application. Additionally, it is unlicensed in most countries, further limiting its use as, given that when unlicensed it's available on a first-come-first-serve basis, reliability could be an issue. The 71–76/81–86 GHz band, referred to as the *E-band* [10] is a licensed band that was approved for use in the United States by the FCC in 2003 and is now approved for use in most countries. Its technology is maturing, and it is becoming very popular as it offers a wide spectrum and provides wide channels, with a data rate over a 2 GHz channel in the 10 Gb/s range for single-polarization transmission and in the 20 Gb/s range for dual-polarization transmission. There are several differing regulations with channel bandwidth varying from about 125 MHz to 2000 MHz, with multiples of 250 MHz being very popular. Next higher in frequency is the licensed *W-Band*, ranging from 92 to 114 GHz. Finally, there is the licensed *D-Band* covering 130–175 GHz. Equipment to operate in these last two bands are in the development stage and when available will likely offer channel bandwidths from 250 MHz up to, where available, 4000 MHz [3]. The major limitation to the use of these non-traditional millimeter-wave bands is short range due to high propagation loss and high atmospheric loss. In the E-band, high-reliability links are typically no longer than a very

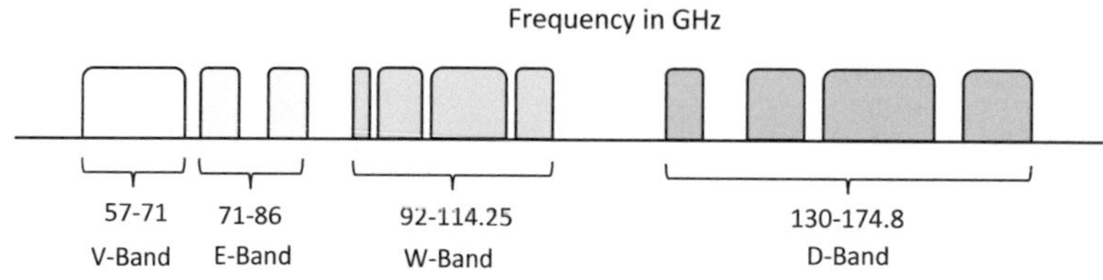

Fig. 1.6 Nontraditional millimeter-wave bands

Table 1.3 Some typical characteristics of nontraditional millimeter-wave bands

Frequency Band	Typical Max. Bandwidth	Cell Location	Typical max. Data rate (Dual pol.)
V: 57–71 GHz	500 MHz	Urban/dense urban	~7 Gb/s
E: 71–87 GHz	2 GHz	Urban/dense urban	~20 Gb/s ~25 Gb/s (future)
W: 92–114 GHz	4 GHz (future)	Urban/dense urban	~50 Gb/s (future)
D: 130–175 GHz	4 GHz (future)	Urban/dense urban	~50 Gb/s (future)

few kilometers. Further, beams are narrower than those in traditional bands and thus call for more stringent transmitter/receiver antenna alignment and very stable antenna mounting structures.

Table 1.3 shows, for the nontraditional millimeter-wave bands, typical bandwidth, cell location, and typical maximum data rates per RF channel.

1.5.3.4 Integrated Access and Backhaul

A key innovation in the evolution of 5G mobile networks is *Integrated Access and Backhaul* (IAB) [11], a technology that is now standardized in 3GPP 5G NR Release 16. It permits the existing 5G access radio air interface technology to be employed in the backhaul network, utilizing access spectrum for backhaul as well. IAB efficiently integrates access and backhaul in the time, frequency, and/or space domain and supports multi-hop backhauling. It can be in-band, where access and backhaul fully or partially share resources in the frequency domain, or it can be out-of-band where access and backhaul don't overlap in the frequency domain. In in-band operation, it can permit a more flexible use of spectrum via the dynamic sharing of it between access and backhaul as compared to a static allocation to both. However, this in-band sharing could impact network quality as a result of interference or could result in a reduction in access capacity.

Though IAB can be supported in both the FR1 (sub 7 GHz) and FR2 (millimeter-wave) 5G NR access bands, the large amount of spectrum available in FR2

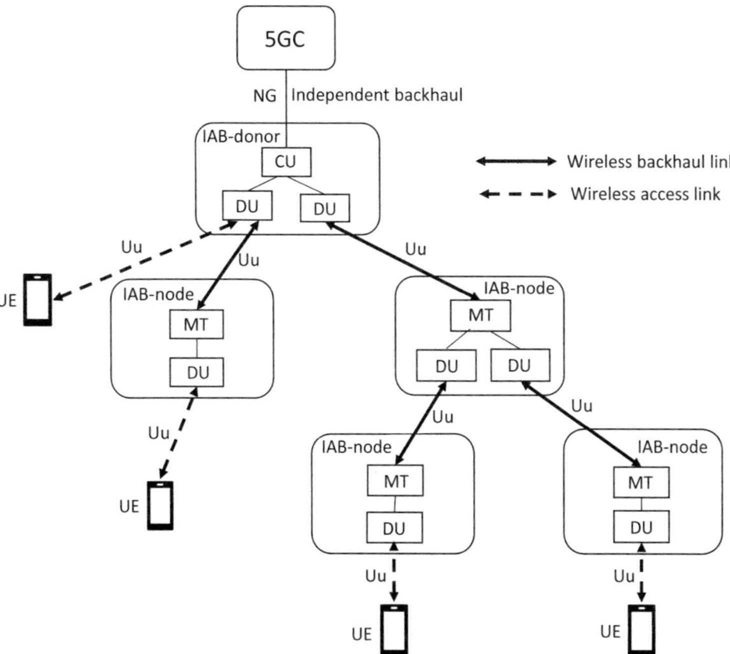

Fig. 1.7 Typical 5G NR Stand Alone IAB Architecture

millimeter-wave bands make them particularly attractive for IAB. Operation in these bands is very supportive of small cell, dense deployments, where there is no existing fiber infrastructure and a cost-effective rapidly deployed backhaul network is desirable.

Figure 1.7 shows typical 5G NR stand-alone (not also connected to the 4G core) IAB architecture. Key components of this architecture are the IAB donor and the IAB node. As shown in the figure, the IAB donor is a gNB that connects to the core network via an NG link and can provide access to UEs directly as well as via IAB nodes. IAB donor connections to UEs and IAB nodes are via the standard 5G NR over-the-air interface, i.e., the Uu interface. The links between the IAB donor and IAB nodes and between individual IAB nodes are wireless backhaul links whereas those between UEs and either the IAB donor or IAB nodes are wireless access links. The IAB node consists of a Mobile Termination (MT) that's interconnected to a DU, the DU being as described in Sect. 1.3 above. The MT is a new entity whose function is to terminate the radio interface layers of the backhaul Uu interface towards the IAB donor or another IAB node. Both the IAB donor and the IAB node may contain multiple DUs. We note that the NG link between the IAB donor and the 5GC is not provided as a part of IAB but rather is provided via an independent fiber or wireless backhaul network.

1.6   Key Wireless Transport Technologies in Support of 5G Networks

A very large number of technologies are required to realize fully functioning 5G network wireless transport links. Some of these technologies, however, are more key than others in the realization of the features and capabilities called for in the 5G world. Following is a short description of these technologies which will be described in some detail in the following chapters:

- *Multi-layer header compression*: As indicated above, communication between the 5G core network and the radio access network is via Ethernet/IP packet data technology. This technology involves multiple layers of header data added to the basic payload for effective communications. However, not all this header data is required when the data is transported over point-to-point wireless links. By compressing this header data and thus allowing more payload per data bit throughput efficiency is increased. Multi-layer header compression is discussed in Chap. 2.
- *Quadrature amplitude modulation* (QAM): Though there are many digital modulation methods, QAM is by far the most popular when high spectral efficiency is required as is the case with 5G wireless transport. Early QAM systems provided realizable spectral efficiency of up about 5 bits per second per Hertz (b/s/Hz) of occupied bandwidth. Some systems today, however, provide a realizable spectral efficiency in excess of 11 b/s/Hz via a transmitted 8192-point signal space constellation diagram and in excess of 12 b/s/Hz via a 16,384-point constellation diagram! QAM modulation is described in Chap. 4.
- *Carrier recovery and timing recovery*: In order to successfully demodulate QAM signals it is necessary to accurately recreate in the receiver the carrier frequency and the timing of the transmitted symbols. Circuitry to accomplish this is described in Chap. 4.
- *Coding*: Real-world digital transmission systems invariable contain imperfections which result, if not addressed, in bit error rate performance that's less than ideal. Coding is a well-established approach which adds redundant bits to wanted bits in such a way as to allow mitigation to a significant degree of this performance degradation as well as improving performance in the presence of thermal noise. One of the highly effective coding techniques that has found favor in 5G wireless transport systems is *low-density parity check* (LDPC) *coding*. A much older coding scheme, *Reed-Solomon coding*, is also often used in these systems. The very latest scheme, *polar coding*, has not yet found application in wireless transport links but is highly effective and may thus be employed in the future. These three coding schemes are addressed in Chap. 5.
- *Transmitter power amplifier predistortion*: A QAM system requires linear transmission. As output power increases, all 'linear' power amplifiers eventually enter a nonlinear compression stage where the signal is distorted. Predistorters seek to distort the signal prior to amplification such that when added to the amplifier's

distortion leads to a close to an undistorted signal. Predistortion is discussed in Chap. 5.

– *Phase noise suppression*: All wireless signals are accompanied by a certain amount of *phase noise* round about center frequency. Phase noise is a function of carrier frequency. Thus, wireless transport links that operate in the non-traditional millimeter-wave bands exhibit phase noise in excess of that experienced in the traditional bands and, if not compensated for, can lead to performance degradation. Phase noise suppression is addressed in Chap. 5.

– *Quadrature imperfection mitigation*: As indicated above, QAM transmission can result in high spectral efficiency. QAM is in effect a system where two separate streams of data are transmitted on the same carrier but in quadrature to each other. That is, one stream is transmitted on a carrier 90^0 out of phase with the other. For performance close to ideal it is necessary that at the receiver, just prior to detection, the original phase and gain relationships between these streams be maintained. This is accomplished with special circuitry which is discussed in Chap. 5.

– *Adaptive equalization*: In ideal over-the-air transmission, there is no distortion to the channel bandpass characteristics. In many real situations, however, channels are distorted and this distortion can lead to degraded error rate performance. Adaptive equalization seeks to adaptively minimize this distortion and is covered in Chap. 5.

– *Link adaptation*: Wireless paths very often exhibit varying losses and in-band distortion. When varying little from ideal, links over such paths can be operated with the highest designed modulation complexity and with a minimum amount of coding. So doing results in maximum spectral efficiency and hence capacity at the specified bit rate reliability. As such paths vary away from the ideal, bit rate reliability can be maintained by reducing modulation complexity and increasing coding gain. Both these actions reduce spectral efficiency and hence capacity. The automatic adjustment of modulation complexity and coding gain in the face of varying channel conditions is referred to as link adaptation and is addressed in Chap. 5.

– *Co-channel dual polarization* (CCDP): This technique to double spectral efficiency has been in use for decades. Here, two different data streams are transmitted over the same channel but each with a different polarization: one horizontal and one vertical. The limitation to its use has been the possible unacceptable interference of one polarized signal into the other. Technology to minimize the impact of this interference is thus key to its successful deployment. Such technology is addressed in Chap. 6.

– *Line-of-sight Multiple Input Multiple Output* (LoS MIMO): As with co-channel dual polarization, LoS MIMO seeks to improve spectral efficiency by transmitting multiple data streams over the same LoS channel via the use of multiple antennas at each end. LoS MIMO is covered in Chap. 6.

– *Orbital angular momentum* (OAM) *multiplexing*: Electromagnetic waves exhibit orbital angular momentum which allows them to be propagated in a number of modes, each orthogonal to the others. By exploiting this phenomenon, data can be multiplexed onto several modes on the same channel thus increasing the channel capacity. OAM multiplexing is still in the research and development phase but in time could find application in wireless transport systems. OAM is discussed in Chap. 6.

– *Band and carrier aggregation* (BCA): In the quest for ever-increasing 5G wireless transport data capacity, clever schemes have been created to transmit data on either aggregated channels within a given band or on aggregated channels in multiple bands. This aggregation is discussed Chap. 6.

1.7 Summary

In this chapter, we first reviewed stated 5G usage scenarios and top-level requirements. Next, a very high-level description of the overall 5G architecture was provided and the location and role of transport connections in this architecture indicated. This was followed by an outline of the various transport alternatives, these including fiber, free space optics, and wireless. The wireless transport outline included a short recital of its evolution, a description of the applicable frequency bands, and an introduction to Integrated Access and Backhaul. Finally, a listing and short description of the key technologies that support wireless transport and are addressed in this text was given.

References

1. Morais DH (2020) Key 5G physical layer technologies. Springer Nature Switzerland AG, Cham
2. 3GPP (2017) TR38.801: study on radio access technology; Radio access architecture and interfaces, Sophia Antipolis
3. ETSI (2018) GR mWT 012 v1.1.1: 5G wireless backhaul/X-haul, ETSI, Sophia Antipolis
4. CPRI (2019) eCPRI interface specification V2.0: common public radio interface: eCPRI interface specification, Ericsson AB, Huawei Technologies Co. Ltd., NEC Corporation, Nokia
5. IEEE (2018) 1914.3-2018 – IEEE standard for radio over ethernet encapsulations and mappings, New York
6. Skogman V Building efficient fronthaul networks using packet technologies, Ericsson Blog, Sweden
7. Ericsson (2020) Ericsson microwave outlook. Ericsson, Goteborg
8. Morais DH (2004) Fixed broadband wireless communications. Prentice Hall, Upper Saddle River
9. Edstam J (2016) Ericsson technology review: microwave backhaul gets a boost with multiband. Ericsson AB, Stockholm
10. Frecassetti M (2020) E-band: survey on the status of worldwide regulation; white paper no. 37, ETSI, Sophia Antipolis
11. 3GPP (2018) TR 38.874; Study on integrated access and backhaul, Sophia Antipolis, France

Chapter 2
5G Transport Payload: Ethernet-Based Packet-Switched Data

2.1 Introduction

Traditional telephone communication is based on *circuit switching*, where circuits are set up and kept continually open between users until the parties end their communications and release the connection. Traditional data communication, however, is based on *packet switching* [1, 2], where the data is sent in discrete packets, each packet being a sequence of *bytes* (each byte contains 8 bits). Contained within each packet are its sender's address and the address of the intended recipient, and each packet seeks the best available route to its destination. In packet-switched networks, the packets at the receiving end don't necessarily arrive in the order sent, so they must be reordered before further use. Here, unlike circuit switching, a circuit is configured and left open only long enough to transport an addressed packet and then freed up to allow reconfiguration for transporting the next packet, which may be from a different source and addressed to a different destination. Packet data communication systems employ so-called protocols. Here, a protocol is a set of rules that must be followed for two devices to communicate with each other. Protocol rules are implemented in software and firmware and cover (a) data format and coding, (b) control information and error handling, and (c) speed matching and sequencing. Protocols are layered on top of each other to form a communication architecture referred to as a *protocol stack*. Each protocol provides a function or functions required to make data communication possible. Typically, many protocols are used so that overall functioning can be broken down into manageable pieces.

5G networks communicate via the Internet. The Internet communicates with the aid of a suite of software network protocols called *Transmission Control Protocol/Internet Protocol* (TCP/IP) or *User Datagram Protocol/Internet Protocol* (UPD/IP). With TCP/IP and UDP/IP, data to be transferred along with administrative information is structured into a sequence of so-called *datagrams*, each datagram being a sequence of bytes. TCP/IP and UDP/IP work together with a physical data link layer to create an effective packet-switched network for Internet communications. The

D. H. Morais, *5G and Beyond Wireless Transport Technologies*,
https://doi.org/10.1007/978-3-030-74080-1_2

data link layer protocol puts the TCP/IP or UDP/IP datagrams into packets (it's like putting an envelope into another envelope) and is responsible for the reliable transmission of packets over the physical layer. The data link protocol associated with TCP/IP and UDP/IP in the 5G world is the *Ethernet* protocol. In the following subsections, a brief introduction to TCP, UDP, IP, and the Ethernet protocol is presented. Further, since voice- and video-generated data are some of the more commonly transported via the Internet to and from 5G networks, how this data is generated is also reviewed. Ethernet transported data carries a lot of nonuser data in the associated headers leading to capacity inefficiency in wireless backhaul links. This inefficiency can be reduced via the use of header compression. Such compression is therefore discussed.

2.2 TCP/IP

The acronym TCP/IP is commonly used in the broad sense and as applied here refers to a hierarchy of protocols occupying four layers but derives its name from the two main protocols, namely TCP and IP. The protocols are stacked as shown in Fig. 2.1. Following is a short description of each protocol layer:

2.2.1 Application Layer Protocol

This is where the user interfaces with the network and includes all the processes that involve user interaction. Protocols at this level include those for facilitating World Wide Web (WWW) access, e-mail, etc. Data from this level to be sent to a remote address is passed on to the next lower layer, the transport layer, in the form of a stream of 8-bit bytes. In the receiving mode, data from the transport layer is passed up to the application layer.

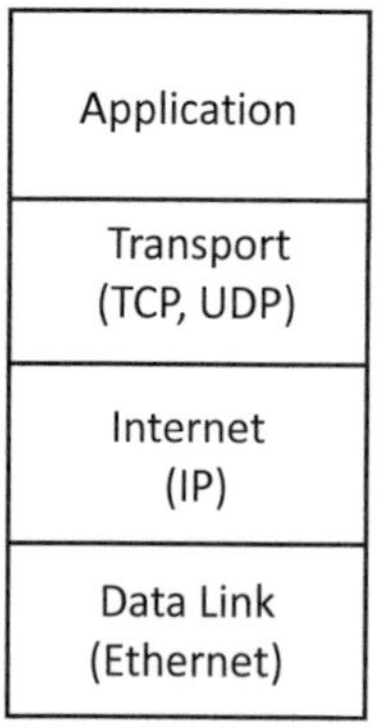

Fig. 2.1 TCP/IP protocol stack

Application
Transport (TCP, UDP)
Internet (IP)
Data Link (Ethernet)

2.2.2   Transport Layer Transmission Control Protocol

As seen in Fig. 2.1, the *Transmission Control Protocol* (TCP) is one of two protocols at the transport layer. TCP provides reliable end-to-end communication and is used by most Internet applications. With TCP, two users, one at each end of the network, must establish a two-way connection before any data can be transferred between them. It is thus referred to as a *connection-oriented* protocol. It guarantees that data sent is not only received at the far end but correctly so. It does this by having the far end acknowledge receipt of the data. If the sender receives no acknowledgment within a specified time frame it resends the data. When TCP receives data from the application layer above for transmission to a remote address, it first splits the data into manageable blocks based on its knowledge of how large a block the network can handle. It then adds control information to the front of each data block. This control information is called a *header* and the addition of the header to the data block is called *encapsulation*. TCP adds a 20-byte header to the front of each data block to form a TCP datagram. It then passes these datagrams to the next layer below, the IP layer. A TCP datagram is shown in Fig. 2.2. When TCP receives a datagram from a remote address via the IP layer below, the opposite procedure to that involved in creating a datagram takes place, i.e., the header is removed, and the data passed to the application layer above. However, before passing it above, TCP takes data that arrives out of sequence and puts it back into the order in which it was sent.

2.2.3   Transport Layer User Datagram Protocol

The *User Datagram Protocol* (UPD) is the second of two protocols at the Transport layer. Unlike TCP, with UDP, users at each end of the network need not establish a connection before any data can be transferred between them. It is therefore described as a *connectionless* protocol. It provides unreliable service with no guarantee of delivery. Packets may be dropped, delayed, or may arrive out of sequence. The

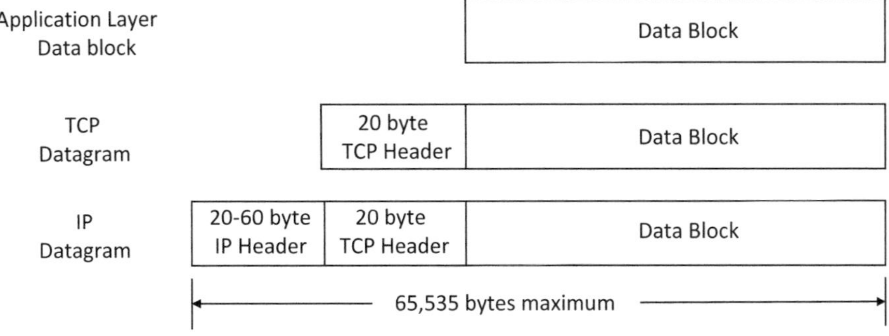

Fig. 2.2 TCP and IP encapsulation

service it provides is referred to as a best-effort one as all attempts are made to deliver packets, with poor reliability caused only by hardware faults or exhausted resources and the fact that no retransmission is requested in the event of error or loss. UDP adds an 8-byte header to the front of each data block to form a UPD datagram. UPD packets arrive more quickly than TCP ones and are processed faster at the cost of potential error. UPD is used primarily for connections where low latency is important, and some data loss is tolerable. Good examples of such connections are voice and video.

2.2.4   *Internet Layer Protocol*

The protocol at this layer, the *Internet Protocol* (IP), is the core protocol of TCP/IP. Its job is to route data across and between networks. When sending datagrams, it figures out how to get them to their destination; when receiving datagrams, it figures out to whom they belong. It is an unreliable protocol, unconcerned as to whether datagrams it sent arrive at their destination and whether datagrams it received arrive in the order sent. If a datagram arrives with any problems, it simply discards it. It leaves the quest for reliable communication, if required, to the TCP level above. Like UDP, it is defined as a *connectionless* protocol. There are, naturally, provisions to create connections per se or communication would be impossible. However, such connections are established on a datagram-by-datagram basis, with no relationship to each other. It processes each datagram as an entity, independent of any other datagram that may have preceded it or may follow it. In fact, there is no information in an IP datagram to identify it as part of a sequence or as belonging to a particular task. Thus, for IP to accomplish its assigned task, each IP datagram must contain complete addressing information. An IP datagram is created by adding a header to the transport layer datagram received from above.

Internet protocols are developed by the *Internet Engineering Task Force* (IETF). *IPv4* was the fourth iteration of IP developed by the IETF but the first version to be widely deployed. It is the dominant network layer protocol on the Internet. Its header is normally 20 bytes long but can in rare circumstances be as large as 60 bytes. A later IP version is IPv6. Its header is 40 bytes long, but 'extension' headers can be added when needed.

An IPv4 datagram created from a TCP datagram is shown in Fig. 2.2. This header includes source and destination address information as well as other control information. As the minimum amount of data transportable is 1 byte, then the minimum size of a data-carrying TCP derived IP datagram is 41 bytes. Its maximum size is specified as 65,535 bytes. All IP networks must be able to handle such IP datagrams of at least 576 bytes in length. IP passes datagrams destined to a remote address to the next layer below, the link layer. On the receiving end, datagrams received by the IP layer from the link layer are stripped of their IP headers and passed up to the transport layer. The maximum amount of data that a link layer packet can carry is

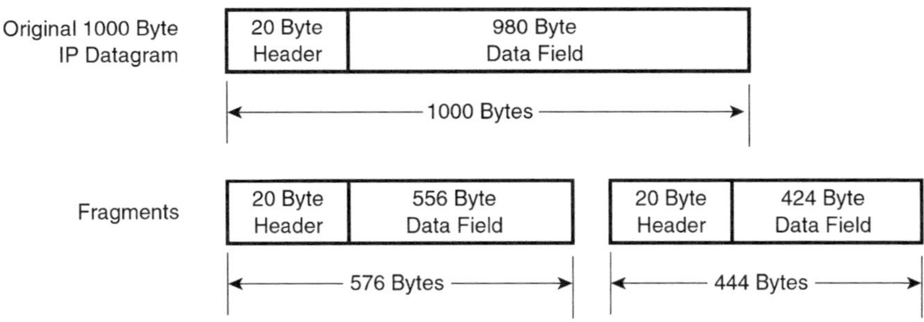

Fig. 2.3 An example of the fragmentation of an IPv4 datagram

called the *maximum transfer unit* (MTU) of the layer. Because, as is explained below, each IP datagram is encapsulated within a link layer packet prior to transmission to the remote address, the MTU of the link layer places a maximum on the length of an IP datagram that it can process. Should an IP datagram be larger than this maximum, it is broken up into smaller datagrams that can fit in the link layer packet. This breaking up process is called *fragmentation* and each of the smaller datagrams created is called a *fragment*. Figure 2.3 shows the fragmentation of an IP datagram 1000 bytes long into two fragments, in order that it may be processed by a link layer with an MTU of 576 bytes. It will be noticed that each fragment has the same basic structure as an IP datagram. Information in the fragment header defines it as a fragment, not a datagram. At the destination, fragments are reassembled to the original IP datagram before being passed to the transport layer. Should one or more of the fragments fail to arrive, the datagram is assumed to be lost and nothing is passed to the transport layer. Fragmentation puts additional tasks on Internet hardware and so it is desirable to keep it to a minimum. Obviously, it can be entirely eliminated by using IP datagrams no longer than 576 bytes since all IP handling networks must have an MTU of at least 576.

For an IP datagram created from a UDP datagram carrying data, the minimum data size is 1 byte. As the UDP header is only 8 bytes long and the IPv4 header can be as little as 20 bytes, a UDP-generated IPv4 data-carrying datagram can be as small as 29 bytes. As with a TCP-generated IP datagram, the maximum size of a UDP-derived IP datagram is 65,535 bytes.

IPv4 uses 32-bit addresses which limits address space to 4295 million possible unique addresses. As addresses available are consumed it appears that an IPv4 address shortage is inevitable. Further, IPv4 provides only limited Quality of Service (QoS) capability. It is the limitation in address space and QoS capability of IPv4 that helped stimulate the push to *IPv6* which is the only other standard internet network layer used on the Internet. IPv6 has a 40-byte header and uses 128-bit addresses which results in approximately 300 billion, billion, billion, billion unique addresses! Further, IPv6 provides for true QoS. IPv4 will be supported alongside IPv6 for the foreseeable future.

An IPv4-to-IPv6 transition technology that involves the presence of IPv4 and IPv6 operating in parallel in an operating system is called *dual-stack IPv4/IPv6*. It enables the application to choose which of the two IP protocols to use or automatically selects it according to address type.

2.2.5 *Data Link Layer Ethernet Protocol*

The data link layer consists of the network hardware that affects actual communication. It stipulates the details of how data is physically sent over the network, including how bits are electrically or optically handled by devices that interface directly with a network link such as twisted-pair copper wire, coaxial cable, optical fiber, or wireless. When transmitting IP data onto the network, it creates packets by taking IP datagrams and adding headers and, in some instances, trailers (control data trailing the datagrams) onto them and dispatches these packets out onto the network. When data packets are received from the network, it strips them of their headers and trailers, if any, and passes them on to the IP layer above. As indicated above, the data link layer protocol associated with TCP/IP in the 5G world is the Ethernet protocol.

IEEE 802.3 standard defined Ethernet packets range in size from 72 to 1526 bytes. Figure 2.4 shows the Ethernet packet format. It consists of a 22-byte header, followed by information data, and ending in a 4-byte trailer.

The header commences with a 7-byte preamble used for synchronization of the receiving station's clock, followed by a 1-byte start frame delimiter used to indicate the start of the frame. This is followed by two 6-byte address codes, the first representing the packet's destination, the second representing the packet's source. Next is a 2-byte length field used to indicate the length of the information data field. Following this is the information data field, which is an integer number of bytes, from a minimum of 46 to a maximum of 1500. Note that if the actual number of data bytes to be sent is less than 46, then extra bytes are added at the end to total 46.

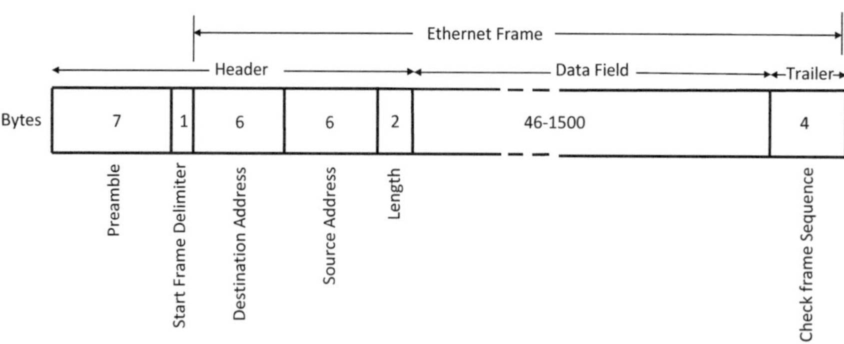

Fig. 2.4 Ethernet packet format

The 4-byte trailer follows the information data field and forms the frame check sequence, which provides the error detection at the bit level to ensure that the data from the destination address through the information data field reaches its destination correctly. The frame check sequence is in fact what is referred to as the checksum resulting from *cyclic redundancy check* (CRC) error detection encoding. Cyclic redundancy check (CRC) [3] is one of the most popular methods of error detection. To create a CRC codeword, a mathematical calculation is carried out on a block of useful binary data which is referred to as a *Frame*. In the calculation, a string of n 0 s is appended to the Frame. This number is divided by another binary number derived from the *generator polynomial*. The number n is one less than the number of bits in the divisor. The remainder of this division represents the content of the Frame and is added to the Frame. These added bits are called the *checksum*. At the receiver, the entire received codeword, i.e., the Frame plus the checksum, is divided by the same polynomial derived bits used by the transmitter. If the result of this division is zero, then the received Frame is assumed to be error-free; if not, an error or errors is assumed to have occurred.

Ethernet devices in computer networks must allow a minimum pause between transmissions of Ethernet packets. This pause is known as the Interframe Gap (IFG) and is necessary to allow for receiver clock recovery, permitting the receiver to prepare for the reception of the next packet. The standard minimum IFG is 96 bits (12 bytes).

The *Ethernet Frame*, as indicated in Fig. 2.4, is defined as the Ethernet packet less the preamble and start frame delimiter. It thus has a minimum size of 64 bytes (14-byte header, 46-byte data, and 4-byte trailer) and a maximum size of 1518 bytes. Despite this maximum IEEE defined frame size, Ethernet networks can accommodate "jumbo" frames that are larger than 1518 bytes and not part of the IEEE standard. Jumbo frame size is vendor-specific, typically includes about 9000 bytes of payload, and most Ethernet equipment can support jumbo frames up to 9216 bytes long.

Ethernet transmission speeds currently in use, defined by the IEEE 802.3 task force, vary from 10 Mb/s all the way up to 100 Gb/s. Ethernet at a speed of 10 Mb/s is referred to simply as *Ethernet*, whereas that at a speed of 100 Mb/s is referred to as *Fast Ethernet* and that at a speed of 1000 Mb/s (1 Gb/s) is referred to as *Gigabit Ethernet*. Above 1 Gb/s the rates defined are 10, 25, 40, 50, and 100 Gb/s, and those defined at 10 Gb/s and above are defined only for full-duplex point-to-point links. With 4G networks, Ethernet transport over any significant distance is normally accomplished with fiber or wireless links and the speed typically employed varies from about 100 Mb/s up to 10 Gb/s. With 5G networks, the Ethernet transport over any significant distance will, like 4G, be normally accomplished with fiber or wireless, but the speed likely to be employed will vary from a low of about 1 Gb/s for rural networks up to in the region of 100 Gb/s for high density-urban networks.

Ethernet over twisted-pair technologies include:

– 10BASE-T, data rates of up to 10 Mb/s, reach up to 100 m, Cat 3 cable
– 100BASE-TX, data rates of up to 100 Mb/s, reach up to 100 m, Cat 5 cable

- 1000BASE-T, data rates of up to 1 Gb/s, reach up to 100 m, Cat 5 cable
- 2.5 GBASE-T, data rates of up to 2.5 Gb/s, reach up to 100 m, Cat 5e cable
- 5GBASE-T, data rates of up to 5 Gb/s, reach up to 100 m, Cat 6 cable
- 10GBASE-T, data rates of up to 10 Gb/s, reach up to 100 m, Cat 6A cable
- 25GBASE-T, data rates of up to 25 Gb/s, reach up to 30 m, Cat 8 cable
- 40GBASE-T, data rates of up to 40 Gb/s, reach up to 30 m, Cat 8 cable

The most common connector used for transmission with these standards is the RJ-45, which is an 8-position, 8-contact (8P8C) modular plug and jack.

Many options exist for Ethernet over optical cable. For 1 Gb/s, these options include:

- 1000BASE-SX, reach up to 1 km
- 1000BASE-LX, reach up to 5 km
- 1000BASE-BX10, reach up to 10 km
- 1000BASE-ZX, reach up to 70 km

For 10 Gb/s, these options include:

- 10GBASE-LR, reach up to 10 km
- 10GBASE-ER, reach up to 40 km
- 10GBASE-ZR, reach up to 80 km

For 25 Gb/s, these options include:

- 25GBASE-SR, reach up to 100 m
- 25GBASE-LR, reach up to 10 km
- 25GBASE-ER, reach up to 40 km

An alternative to passive Ethernet connection via twisted-pair copper cable and RJ-45 connections is an active connection using *Small Form Factor Pluggable* (SFP) technology and evolutions thereof. With SFP, the equipment to be connected is equipped with an SFP port, also known as an SFP slot. Into this port is inserted a compact, hot-pluggable transceiver that in turn is connected to a fiber optic cable or sometimes a twisted pair via RJ-45. There are many different types of optical cable connectors, but SPF modules typically use the LC connector.

All the 1 Gb/s Ethernet over optical cable options shown above are transportable via SFP transceivers. For the 10 Gb/s options, an enhanced form of SFP, called SFP +, but with the same physical dimensions of SFP, is used. For the 25 Gb/s options, the SFP28 transceiver is used, which like the SFP+ has the same form factor as the SFP.

A slightly larger version of SFP/SFP+/SFP28 is the four-lane Quad Small Form Factor Pluggable (QSFP/QSFP+/QSFP28, all of the same form factor) which allows for speeds four times the corresponding SFP, SFP+ and SFP28 speeds, i.e., 4 Gb/s, 40 Gb/s, and 100 Gb/s, respectively.

Ethernet Header 22-byte	**MPLS Header** **4-byte**	IPv4/IPv6 Header 20/40-byte	Payload	Trailer 4-byte

Fig. 2.5 Ethernet frame encompassing an MPLS header

2.2.6   Multi-Protocol Label Switching (MPLS)

Wireless IP transport networks are typically aided by the application *Multi-Protocol Label Switching* (MPLS). MPLS, developed by the IETF, is a protocol for routing data traffic through a telecommunication network from one router to the next. When applied to Ethernet transported IP data, it is situated between the Internet protocol layer and the Ethernet protocol layer. In a standard IP network, as a packet of data moves from router to router towards its final destination, each router must perform an IP lookup and make an independent decision as to what route the packet should take for the next hop, then forward the packet on to the next router. With MPLS, routing is done via "label switching" instead. Here, the first router does a routing lookup, finds the final destination router, finds a specific path to that router, and then applies a "label" to the packet based on the determined path. The following routers then use the label to route the packet without needing to perform any additional IP lookups. When the packet arrives at its destination router, the label is removed, and the packet is delivered via normal IP routing. The MPLS header is 32 bits (4 bytes) long and is situated within the Ethernet frame as shown in Fig. 2.5.

An original key feature of MPLS was its ability to reduce routing lookup thus reducing the computation required on the involved routers. However, modern highly integrated circuitry has mostly removed this advantage. MPLS still finds favor, however, because it also provides a number of other advantages, including the ability to control where and how traffic is routed on the network and the ability to prioritize different services, for example, providing higher-priority and lower-latency routes to delay intolerant services such as real-time voice and video.

2.3   Voice over IP (VoIP)

In the narrowest sense, *Voice over Internet Protocol* (VoIP) implies voice communication over only an IP link. In the broader and more realistic sense, it means voice communication where at least a part of the network is an IP link. Conceptually, communication over an all-IP VoIP network is quite straightforward:

- The analog voice signal is converted to a digital stream, normally at 64 kb/s.
- This data stream is then typically compressed using a *codec* (a codec is a compression/decompression device).
- Bits in this compressed stream are grouped into packets.
- These packets are transported via an IP network to a destination.

IP Header (20 bytes)	UDP Header (8 bytes)	RTP Header (12 bytes min.)	Real Time Payload, e.g. voice, video

Fig. 2.6 Structure of an IPv4 voice packet

– At the destination, they are converted back into the original digital stream and, finally, back into an analog signal.

The structure of an IPv4-based VoIP packet is shown in Fig. 2.6. Note that immediately ahead of the voice payload data a new header has been added. This header is called the *Real-time Transport Protocol* (RTP) [4] and was developed by the IETF. It's typically 12 bytes long. RTP is a protocol to facilitate the delivery of audio and video over IP networks and typically runs over UDP. It addresses jitter compensation, detection of packet loss, and out-of-order delivery, issues which are likely with UDP transmissions. We note that now the total packet header size is 40 bytes.

As 4G and 5G networks only support packet services, then voice service provided over these networks must be via VoIP. Adaptive Multirate (AMR) codecs are the ones most commonly used for creating VoIP data. AMR codecs come in two versions, namely AMR-Narrowband (AMR-NB) [5] and AMR-Wideband (AMR-WB) [6].

AMR-NB uses a sampling frequency of 8-kilo samples per second (ks/s) and provides an audio bandwidth of 300–3400 Hz. Its output data rates range from 4.75 kb/s to 12.2 kb/s. AMR-WB uses a sampling frequency of 16 ks/s and provides an audio bandwidth of 50–7000 Hz. Its output data rates range from 6.6 kb/s to 28.85 kb/s, with a commonly used rate of 12.65 kb/s. For comparable data rates, AMR-WB offers substantially better voice quality than AMR-NB, which is not surprising, given AMR-WB's much wider audio bandwidth.

The size of the voice payload data sent over VoIP packets depends on the coding rate and the voice sampling period. Typically, the voice sampling period is 20 ms. With a coding rate of 12.65 kb/s, 253 bits are generated every 20 ms. This computes to 31.625 bytes, but as data payload is transmitted in an integer number of bytes this would be transmitted as 32 bytes. Thus, with IPv4, and therefore a packet header size of 40 bytes, the total VoIP packet would be 72 bytes, and the header overhead would be 55% of the total packet. Had we assumed IPv6, then the packet header size would have been 60 bytes leading to a header overhead of 65%.

2.4   Video over IP

Video over IP from a technical point of view is identical to voice over IP with the exception that we are dealing with much higher data rates. These higher rates are the result of the inherently much higher information content of video compared to voice.

Uncompressed video rates are very high and hence much compression is required to achieve manageable transmission rates. Video signal compression is a complex process and will not be addressed here. Suffice it to say, however, that it is possible because there are redundancies in a video signal's uncompressed format and because visual perception has certain characteristics that can be used to advantage. Regarding format redundancies, video compression in essence results in the transmission of only differences from one video frame to the next, rather than the transmission of each new frame in its entirety.

Many video codec standards are available, but to get a sense of their capability we will look at the compression capability of one of the most popular, namely the *H.264 Advanced Video Coding* (H.264/AVC) standard. To do this we will first determine the uncompressed video data rate, VR_{UC} say. VR_{UC} is given by:

$$VR_{UC} = \text{color depth} \times \text{number of vertical pixels}$$
$$\times \text{ number of horizontal pixels} \times \text{refresh rate (frames per second)} \quad (2.1)$$

where:

- Color depth is the number of bits used to indicate the color of a single pixel, and.
- Refresh rate is the rate at which the screen presentation is refreshed.

For the 1080p high definition (HD) video format, with a color depth of 24, 1080 vertical pixels, 1920 horizontal pixels, and a refresh rate of 30, VR_{UC} computes to 1493 Mb/s. Video compression possible with H.264 varies. H.246 offers a number of profiles. Its 'Baseline' profile, for example, can achieve a compression rate of up to about 1000:1. Its 'High' profile can achieve a compression rate of up to about 2000:1. These are impressive numbers, but as compression rate increases picture quality decreases so there is a practical limit to how much compression is normally used. For the 1080p format outlined above, a compressed bit rate of about 5 Mb/s or more is required for minimum acceptable picture quality. Assuming a rate of 5 Mb/s, then the compression rate is 1493/5, or approximately 300. For 4 K Ultra HD (2160 vertical x 3890 horizontal pixels), a compressed bit rate of about 15 Mb/s is required for minimum acceptable quality, and a rate of about 25 Mb/s is required for good quality.

2.5 Header Compression

In data packet transmission, depending on the amount of user data per packet, the quantity of overhead data contained in headers compared to user data can be surprisingly large. This overhead is required for routing, collision avoidance, flow identification, etc. in complex LAN/MAN networks. Wireless packet data transmission capacity is, in general, much more limited than that over fiber. With fiber, the capacity overhead due to headers tends not to impact the ability to transmit the payload capacity. With wireless, this is often not the case. This is particularly so with 5G transport links where often very high useful data rates are required thus stretching the capacity limitation of the links.

In a point-to-point wireless link with full-duplex transmission, the medium is not shared by uncoordinated users simultaneously. This provides several benefits. For example, collisions don't occur, and no overhead structure is thus required to handle this. Another example is that in the Ethernet packet the Interframe gap, the Preamble, and the start frame delimiter are not required and can be stripped from the packet. The remaining headers can then be subject to compression. The overall result is that it is possible to strip away or compress unnecessary header information leading to increased useful data capacity. *Header compression* relies on many fields being unnecessary, being constant, or rarely changing in consecutive packets belonging to the same packet flow. Typically, with header compression, for packets of the same flow with the same destination, the parts of the header information that are unchanging are only sent at the start of the flow and updated after only a defined time or if some change has occurred. At the point of transmission, each unique header is replaced with a unique identity of a much smaller size, and the process is reversed on the receiving size.

Header compression can be accomplished at the Ethernet layer level, for example by the EthHC™ scheme developed by Effnet AB. However, it can also be accomplished at the IP level as well as the TCP/UDP and the RTP level. As was shown in Sect. 2.3, header created overhead in VoIP transmission is very high. Because this overhead is clearly wasteful of capacity, header compression is normally used along with VoIP. This can be accomplished via *Robust Header Compression* (ROHC) [7], a compression technique used in streaming applications such as voice and video. It compresses the IP, UDP, and RTP headers from 40 or 60 bytes typically into only one or three bytes!

Most 5G wireless transport links typically use header compression schemes that support compression of Ethernet, MPLS, IPv4, IPv6, TCP, UDP, and RTP, as, for example, the EffnetBHC™ scheme developed by Effnet AB. As is to be expected, header compression provides higher capacity gain for packets of smaller frame size since here their headers represent a relatively larger portion of the total frame size. Typically, it provides 5–10% gain with Ethernet and IPv4 with an average frame size of 400–600 bytes, and a 15–20% gain with Ethernet, MPLS, and IPv6 with the same average frame size [8].

2.6 Payload Compression

In addition to header compression, compression of the payload can also be used to further reduce the Ethernet stream data rate. *Payload compression* is accomplished by analyzing the traffic data for repetitive bit patterns and then applying this data to algorithms that reduce the number of packets being transmitted. At the received end, the compressed payload is restructured into its original packets. The gain from payload compression is very dependent on the type of data being transported. For uncompressed and/or unencrypted data streams, it can be on the order of tens of percentage points. However, it is much more limited if the data distribution is random as is the case with compressed and encrypted data streams. Since mobile broadband payload traffic is normally compressed and encrypted, compression gain on wireless transport links is typically limited to about 2–3%.

2.7 Summary

5G networks communicate via the internet with the aid of the TCP, UPD, IP, and Ethernet protocols. In this chapter, these protocols were reviewed. Further, since voice- and video-generated data are some of the more commonly transported via wireless to and from 5G networks, how this data is generated was also reviewed. Finally, compression of the headers associated with Ethernet transported data, applied in order to improve capacity efficiency, was discussed.

References

1. Lee BG, Kang M, Lee J (1993) Broadband telecommunications technology. Artech House, Inc., Norwood
2. Bates B, Gregory D (1996) Voice & data communications handbook. Mc-Graw-Hill, Inc., New York
3. Peterson WW, Brown DT (1961, January) Cyclic codes for error detection. In: Proceedings of the IRE, vol 49
4. Schulzrinne H et al. (2003, July) IETF RC 3550 RTP: a transport protocol for real-time applications
5. 3GPP TS 26.071 (2018, June 22) Mandatory speech CODEC speech processing functions; AMR speech CODEC; General description Version 1500
6. 3GPP TS 26.190 (2018, June 22) Speech codec speech processing functions; Adaptive Multi-Rate – Wideband (AMR-WB) speech codec; Transcoding functions Version 1500
7. IETF RFC 5795 (2010, March) The Robust Header Compression (ROHC) framework
8. Ericsson AB (2014) Ericsson microwave towards 2020 report

Chapter 3
The Fixed Wireless Path

3.1 Introduction

A fixed wireless system communicates between sites via the propagation of radio waves over a *path*. In an ideal world, the path would linearly attenuate the transmitted signal by a fixed amount resulting in an undistorted signal at a predictable level at the receiver input. A digital system so deployed in "free space" could then be designed so that the received level resulted in an acceptable bit error rate and no further analysis would be required. In fact, ground-station-to-satellite paths behave close to such an ideal. For point-to-point terrestrial links, however, atmospheric anomalies and the intervening terrain often result in a significant deviation from the ideal postulated above, such deviation being referred to as *fading*. A typical fixed wireless link is shown in Fig. 3.1. As will be seen in succeeding sections, the maximum length of a wireless path for reliable communications varies depending on (a) the propagation frequency; (b) the antenna heights; (c) terrain conditions between the sites, in particular, the minimum clearance between the direct signal path and ground obstructions; (d) atmospheric conditions over the path; and (e) the radio equipment and antenna system electrical parameters. For systems operating in the 6–11 GHz bands, paths are typically up to 30–50 km in length. For those operating in the 18–42 GHz bands, however, path length becomes more and more restricted, the length in kilometers being typically in double digits to low single digit, much of this restriction being due to fading resulting from rain. Finally, for systems operating in the non-traditional millimeter-wave bands (bands above 50 GHz) path lengths are very short, typically up to just a very few kilometers due to fading resulting from rain and other atmospheric effects.

In this chapter, we will examine propagation in an ideal environment and then study the various types of fading. As antennas provide the means to efficiently launch and receive radio waves, a brief review of their characteristics is in order as the first step in addressing ideal propagation.

D. H. Morais, *5G and Beyond Wireless Transport Technologies*,
https://doi.org/10.1007/978-3-030-74080-1_3

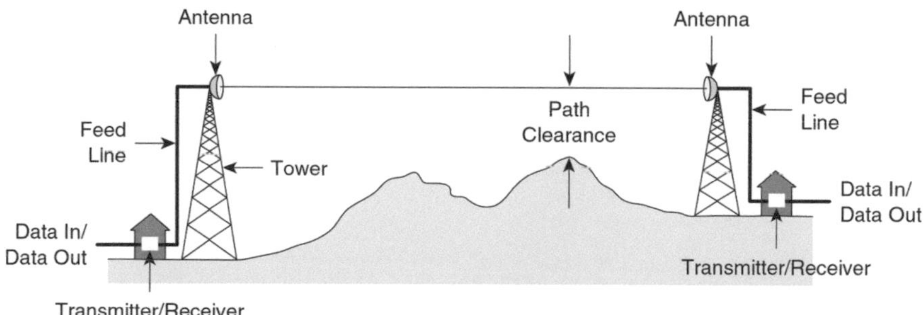

Fig. 3.1 Typical fixed wireless link

3.2   Antennas

3.2.1   Introduction

There are myriads of antenna shapes and designs for wireless communication, often a function of operating frequency. For the fixed wireless systems of interest, operating frequencies are currently between about 6 GHz and 87 GHz (top of the E-band) but in the future can be expected to extend to about 114 GHz (top of the W-band) and later still up to about 170 GHz (top of the D-band). Over the current range of frequencies, only two types of antenna are typically used for point-to-point transport links, namely, the *parabolic antenna* and, more recently, the *flat plane* or *planar array antenna*. As the gain of antennas increase, they become more and more directive. For point-to-point links, this feature is desirable, with a transmitting antenna of high gain focusing its energy in a narrow beam that is directed to a receiving antenna where it is collected in an efficient manner.

Many antenna characteristics are important in designing point-to-point wireless systems. The most important of these will be reviewed followed by a brief description of those antennas commonly used in such systems.

3.2.2   Antenna Characteristics

Antenna gain is the most important antenna characteristic. It is a measure of the antenna's ability to concentrate its energy in a specific direction relative to radiating it isotropically, i.e., equally in all directions. The more concentrated the beam, the higher the gain of the antenna. It can be shown [1] that a transmitting antenna that concentrates its radiated energy within a small beam has a gain G in the direction of maximum intensity with respect to an isotropic radiator of

$$G(dB) = 10 \cdot \log_{10}\left(\frac{4\pi A_{ef}}{\lambda^2}\right) \tag{3.1}$$

where

A_{ef} = effective area of the antenna aperture in (meters)^2
λ = wavelength of the radiated signal in meters.

The antenna's physical area A_p is related to its effective area by the following relationship

$$A_{ef} = \eta A_p \tag{3.2}$$

where η is the efficiency factor of the antenna.

Equation (3.1) indicates that antenna gain is function of the square of the frequency ($\lambda f = c$), thus doubling the frequency increases the gain by 6 dB. It is also a function of the area of the aperture. Thus, for a parabolic antenna, gain is a function of the square of the diameter and so doubling the diameter increases the gain by 6 dB. The efficiency factor η in Eq. (3.2) accounts for the fact that the antenna is not 100% efficient. This is because the total incident power on the radiator is not radiated forward as per theory. Some of it is lost to "spillover" at edges, some is misdirected because the radiator surface is not manufactured perfectly to the desired shape, and some is blocked by the presence of the feed radiator. The operation of an antenna in the receive mode is the inverse of its operation in the transmit mode. It, therefore, comes as no surprise that its receive gain, defined as the energy received by the antenna compared to that received by an isotropic absorber, is identical to its transmit gain.

Figure 3.2 shows a plot of antenna gain versus angular deviation from its direct axis, which is referred to as its *pole* or *boresight*. It shows the main lobe where most of the power is concentrated. It also shows side lobes and back lobes that can cause interference into or from other wireless systems in the vicinity.

The *beamwidth* of an antenna is closely associated with its gain. The higher the gain of the antenna the narrower the width of the beam. Beamwidth is measured in radians or degrees and is usually defined as the angle that subtends the points at which the peak field power is reduced by 3 dB. Figure 3.2 illustrates this definition. Antennas used in wireless transport systems have beamwidths that vary from a fraction of a degree to several degrees. The narrower the beamwidth, the more

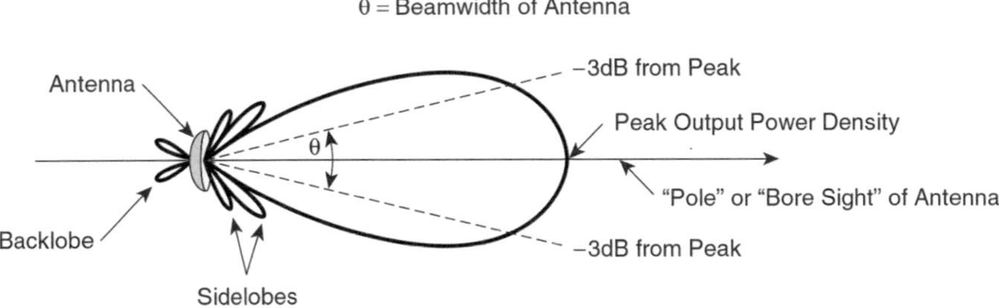

Fig. 3.2 Antenna gain versus angular deviation from its pole

interference from external sources including nearby antennas is minimized. This improvement comes at a price, however, as narrow beamwidth antennas require precise alignment and if of large diameter have a higher wind load, necessitating increased mechanical stability support. This is because a very small shift in alignment results in a measurable decrease in the transmitted power directed at a receiving antenna and in the received power from a distant transmitting antenna.

The *front-to-back ratio* is another important antenna characteristic. It is defined as the ratio, usually expressed in dBs, of its maximum gain in the forward direction to the maximum gain in the region of its backward direction, the latter being the maximum back-lobe gain. This ratio is particularly critical in repeater systems where the same frequencies are used in both directions for a station. Unfortunately, the front-to-back ratio of an antenna in a real installation can vary widely from that in a purely free space environment. This variation is due to foreground reflections in a backward direction of energy from the main transmission lobe by objects in or near the lobe. These backward reflections can reduce the free space only front-to-back ratio by 20 to 30 dB.

The polarization of an antenna refers to the alignment of the electric field in the radiated wave. Thus, in a *horizontally polarized* antenna, the electric field is horizontal, and in a *vertically polarized* antenna, the electric field is vertical. When a signal is transmitted in one polarization, only a small fraction is likely to be converted to the other polarization due to imperfections in the antenna and the path. The ratio of the power received in the desired polarization to the power received in the undesired polarization is referred to as *cross-polarization discrimination*. Cross-polarization discrimination typical varies from about 25 to 40 dB depending on the antennas and the path.

To improve adjacent channel discrimination most standard frequency assignments are such that adjacent channels operate on different polarizations. This way, the adjacent channels interfering levels are attenuated not only via filtering but also via cross-polarization discrimination.

Finally, we note that an antenna-related parameter often regulated for fixed wireless systems is the *Equivalent Isotropically Radiated Power* (*EIRP*). This power is the product of the power supplied to the transmitting antenna, P_t, say, and the gain of the transmitting antenna, G_{ta} say. Thus, we have

$$EIRP = P_t \cdot G_{ta} \tag{3.3}$$

3.2.3 *Typical Point-to-Point Wireless Antennas*

The *parabolic antenna* is the most commonly used type with wireless transport systems. It consists of a parabolic reflector which in the transmit mode is radiated with RF energy from a feed source located at the focus of the reflector. In its simplest form, the parabolic reflector is the only reflector used, and the feed source radiates

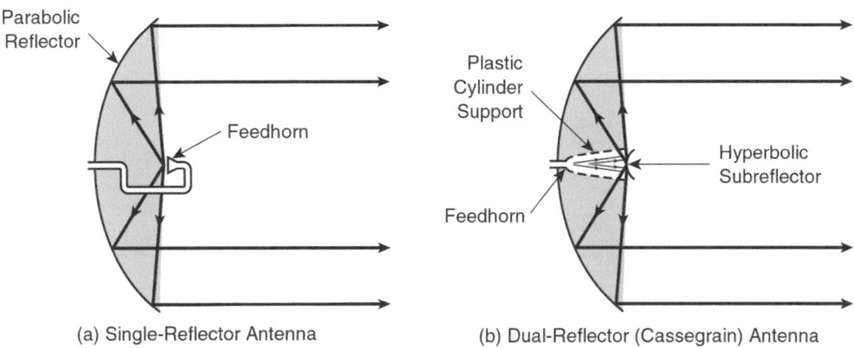

Fig. 3.3 Cross sections of Single- and dual-reflector parabolic antennas

directly onto it. The feed source consists of a waveguide that opens in the form of an enlarging taper in such a way as to match the impedance of the waveguide to that of free space. Such a feed source is, in fact, a member of the horn antenna class and is commonly referred to as the antenna feed horn. A diagram of a cross section of such a parabolic antenna is shown in Fig. 3.3a. As can be seen, the antenna's operation is very similar to that of a flashlight, RF waves following most of the rules of optics. One problem associated with directly radiating feed horns is that their support structure blocks part of the beam they radiate and thus degrade performance. Parabolic antennas vary in diameter, from as low as about 0.2 m for those operating at the high end of the traditional frequency bands (30–42 GHz) and in the E-Band (71–86 GHz), to a high of about 3.7 m for those operating at the low end of the traditional bands (6–11 GHz). The performance of a parabolic antenna radiated directly from a feed horn can be enhanced by adding a *shield*, which takes the form of a forward projecting short cylinder attached to its circumference. The shield is normally covered by non-radiation absorbing material called a *radome*. A shielded antenna will typically have slightly higher gain, slightly narrower beamwidth, and measurably higher (in the range of 10–20 dB) front-to-back ratio than its unshielded counterpart. Another version of the parabolic antenna that achieves these enhanced performance characteristics without the use of a shield is a dual-reflector type as shown in Fig. 3.3b. Here the feed horn is located at the apex of the main reflector and radiates a hyperbolic subreflector that in turn radiates the main reflector. Such a dual reflector type is called a *Cassegrain* antenna. The subreflector is normally rigidly supported by a special plastic cylinder and so there is minimum radiated beam blockage.

The nominal value of η, the efficiency factor for parabolic antennas is approximately 0.55 (55%). From Eqs. (3.1) and (3.2) and assuming η equals 0.55, then for a parabolic antenna, gain G in dBs can be shown to be given by:

$$G = 20Log_{10}D + 20Log_{10}F + 17.8 \tag{3.4}$$

where D is the diameter in meters and F is the frequency in GHz.

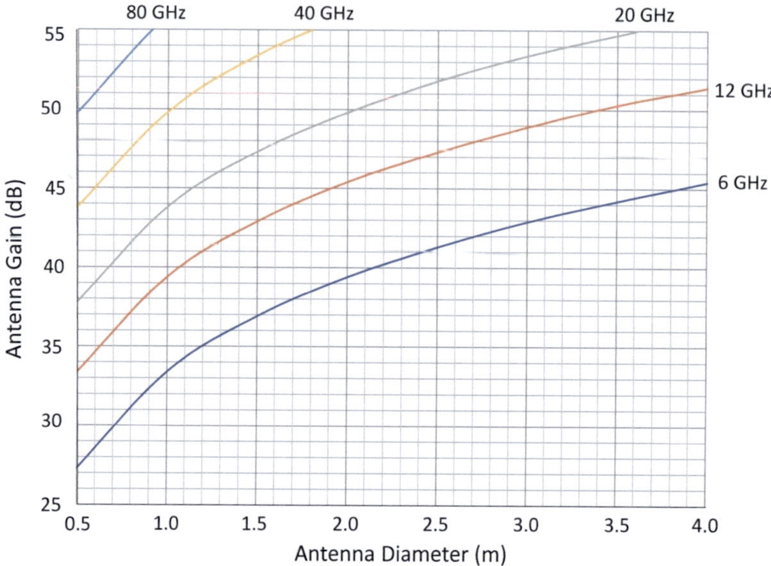

Fig. 3.4 Parabolic antenna gain versus diameter and frequency

Figure 3.4 shows antenna gain versus diameter and frequency for parabolic antennas assuming gain is as given in Eq. (3.4).

The beamwidth in degrees, θ, of a parabolic antenna is given approximately by $70\lambda/D$ where λ is the wavelength in meters and D is the diameter of the antenna in meters. This relationship can be restated as:

$$\theta \approx 21/Df \tag{3.5}$$

where D is the diameter in meters and f is the frequency in GHz.

Thus, we see that it is inversely proportional to both the frequency and antenna diameter and therefore in turn inversely proportional to the square root of the absolute gain. For an antenna with a diameter of 2 m operating at 6 GHz and hence with a gain of approximately 40 dB, the beamwidth computes to 1.75°. For an antenna with a diameter of 0.6 m operating at 80 GHz and hence with a gain of approximately 51.4 dB the beamwidth computes to 0.44°. Proper alignment with such a narrow beamwidth is achievable but somewhat difficult. Thus, as a practical rule, maximum antenna gain in dBs is normally limited to the low 50s, as with values in excess of this proper alignment becomes very challenging.

A single parabolic antenna, in addition to being able to transport a single-polarized signal, can also be made to transport both a vertical as well as a horizontally polarized signal simultaneously. Such antennas are referred to as dual-polarized ones. Dual polarization transport is achieved either by the use of dual feed horns or via the use of an *orthomode transducer* (OMT), also called a polarization duplexer.

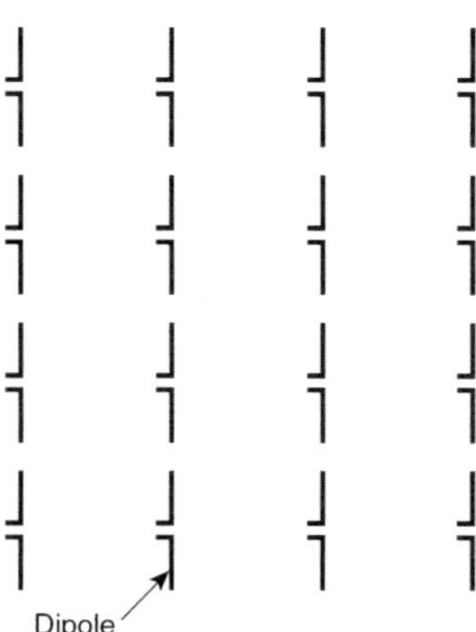

Fig. 3.5 Typical dipole planar array

Rectangular waveguides and feed horns support only one polarization whereas square or circular waveguides and feed horns support dual-polarized signals. In the transmit direction, the OMT thus accepts via rectangular flanges the two orthogonally polarized signals, combines them, and feeds them out via a square or circular flange to a square or circular feed horn. In the receive direction, it divides the two received polarized signals feeding them to the separate rectangular flanges.

The *flat panel antenna* or *planar array antenna* is a directive antenna whose key attribute is its flat profile. Its use has been spurred by the need to provide an environmentally unobtrusive presence in many metropolitan areas. Unfortunately, the pleasing aesthetics of the planar array typically comes at the expense of lower gain and hence lower directivity relative to a parabolic antenna of equal aperture.

One physical realization of a flat panel antenna is one where several dipole antennas are arranged in a planar array, for example, as shown in Fig. 3.5. Such an arrangement provides two dimensions of control, allowing a beam, directive in both the horizontal and vertical coordinates, to be produced. Though each dipole in the array has an omnidirectional radiation pattern, they are cleverly connected together via a network that results in directivity in a forward direction, normally along a line perpendicular to the plane of the array.

Another physical realization of the flat panel antenna is the *patch array antenna* which consists of "patches" arranged in a planar array. Each patch is a rectangular sheet of metal, approximately half the operating frequency wavelength on each side, mounted over, but electrically insulated from, a sheet of metal serving as a ground plane. The feed point is along one edge of the rectangle, and for cross-polarized operation, two feed points are located on perpendicular sides. Figure 3.6 is a graphical representation of a 2x2 patch array antenna. Millimeter-wave flat panel antennas are usually patch antennas.

We note that the gain and hence directivity of a linear array in the horizontal or vertical plane is a direct function of the number of antenna elements in that plane. Thus, the array shown in Fig. 3.5 that has four elements on the horizontal axis has a gain in the horizontal plane four times, i.e., 6 dB above, that of a single dipole. The same applies to the vertical plane.

Fig. 3.6 2x2 Patch Array Antenna

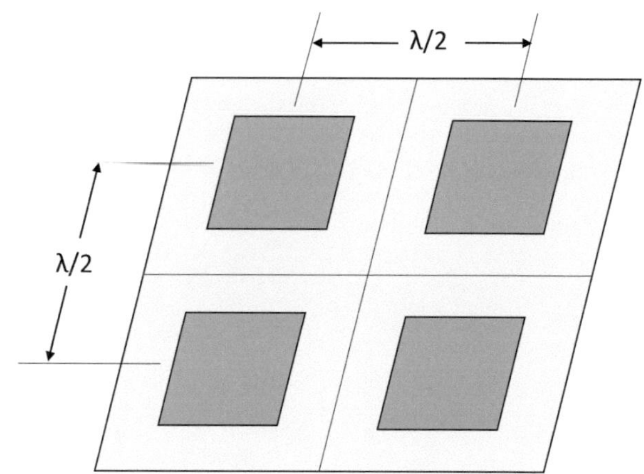

3.3 Free Space Propagation

As indicated above, the propagation of a signal over a wireless path is affected by both atmospheric anomalies and the intervening terrain. Absent any such interfering effects, we have a signal loss only as a result of free space. *Free space loss* is defined as the loss between two isotropic antennas in free space.

Consider the point source shown in Fig. 3.7 radiating isotropically a signal of power P_t into free space. As the surface area of a sphere of radius d is $4\pi d^2$, then the radiated power density $p(d)$ on a sphere of radius d, at any point on the sphere, is given by

$$p(d) = \frac{P_t}{4\pi d^2} \tag{3.6}$$

If a receiving antenna with an effective area A_{ef} is located on the surface of the sphere, then the power received by this antenna, P_r, will be equal to the power density on the sphere times the effective area of the antenna, i.e.,

$$P_r = \frac{P_t A_{ef}}{4\pi d^2} \tag{3.7}$$

To determine free space loss we need to know the power received by an isotropic antenna. But, in order to know this, Eq. (3.7) indicates that we need to know the effective area of an isotropic antenna. By definition, the gain G of an isotropic antenna is 1. Substituting this value of gain into Eq. (3.1) gives

$$A_{ef} = \frac{\lambda^2}{4\pi} \tag{3.8}$$

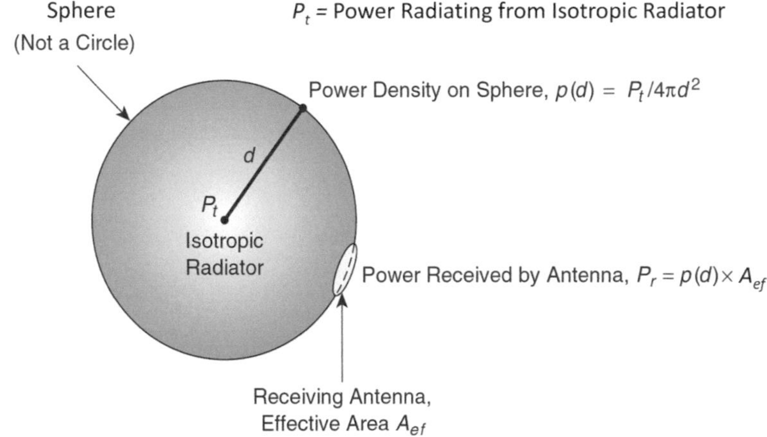

Fig. 3.7 Power density radiated from an isotropic radiator and power received by an antenna

Substituting Eq. (3.8) into Eq. (3.7) we determine that P_r, the power receiver by an isotropic antenna is related to P_t, the power transmitted by an isotropic radiator by the following relationship

$$P_r = \frac{P_t}{\left(\frac{4\pi d}{\lambda}\right)^2} \tag{3.9}$$

The denominator of the right-hand side of Eq. (3.9) represents the free space loss, L_{fs}, experienced between the isotropic antennas. It is usually expressed in its logarithmic form, i.e.,

$$L_{fs}(dB) = 20\log_{10}\left(\frac{4\pi d}{\lambda}\right) \tag{3.10}$$

Substituting in Eq. (3.10) the well-known relationship $\lambda f = c$, where c is the speed of electromagnetic propagation and equals 3×10^8 m/s for free space transmission, we get

$$L_{fs}(dB) = 92.4 + 20\log_{10}f + 20\log_{10}d \tag{3.11a}$$

where

f is the transmission frequency in GHz and.
d is the transmission distance in km,

and

$$L_{fs}(dB) = 96.5 + 20\log_{10}f + 20\log_{10}d \tag{3.11b}$$

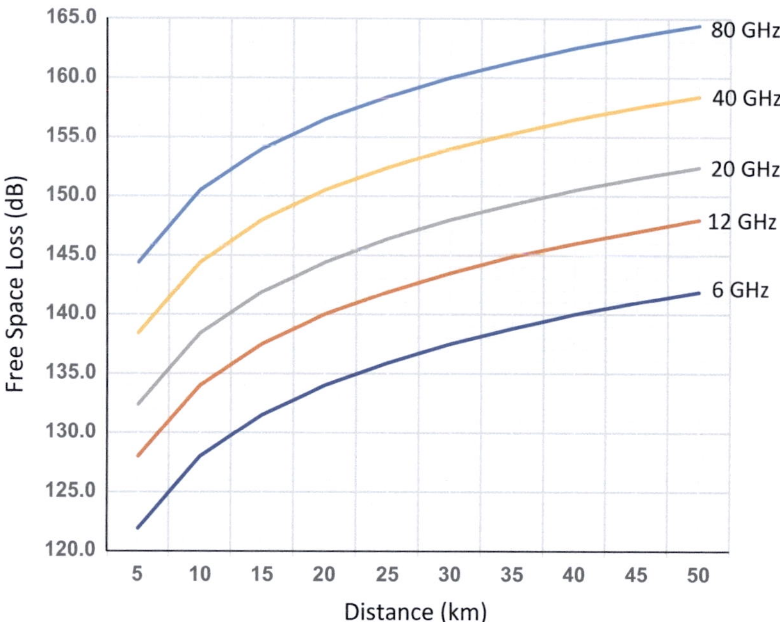

Fig. 3.8 Free space loss versus distance and frequency

where

f is the transmission frequency in GHz and,
d is the transmission distance in miles.

Figure 3.8 shows free space loss versus distance and frequency. We note from the figure that for a 6 GHz path 50 km long the free space loss is approximately 142 dB and that for a 40 GHz path with the same loss the path length is only 8 km.

3.4   Line-of-Sight Non-Faded Received Signal Level

Given the relationship to determine free space loss one is now able to also set out a relationship for the determination of the receiver input power P_r in a typical wireless link as shown in Fig. 3.1, assuming no loss to fading or obstruction. Starting with the transmitter output power P_t at the transmitter antenna port one simply accounts for all the gains and losses between the transmitter output and the receiver input. These gains and losses, in dBs, in the order incurred are:

L_{tf} = Loss in transmitter antenna feeder line, if any (coaxial cable or waveguide depending on frequency)
G_{ta} = Gain of the transmitter antenna
L_{fs} = Free space loss

L_a = Free space absorption, if any
G_{ra} = Gain of the receiver antenna
L_{rf} = Loss in the receiver antenna feeder line, if any

Thus

$$P_r = P_t - L_{tf} + G_{ta} - L_{fs} - L_a + G_{ra} - L_{rf} \qquad (3.12a)$$

$$= P_t - L_{sl} \qquad (3.12b)$$

where $L_{sl} = L_{tf} - G_{ta} + L_{fs} + L_a - G_{ra} + L_{rf}$ is referred to as the *section loss* or *net path loss*.

Example 3.1 Computation of Received Input Power
A microwave link has the following parameters:

Path length, d = 40 km.
Operating frequency, f = 6 GHz.
Transmitter output power, P_t = 30.0 dBm (1 Watt).
Loss in transmitter antenna feeder line, L_{tf} = 3 dB.
Transmitter antenna gain, G_{ta} = 40.5 dB.
Free space absorption, L_a = 0.5 dB.
Receiver antenna gain, G_{ra} = 37 dB.
Loss in receiver antenna feeder line, L_{rf} = 2.5 dB.

What is the receiver input power?

Solution
By Eq. (3.11a) the free space loss, L_{fs}, is given by

$$L_{fs} = 92.4 + 20\log_{10}6 + 20\log_{10}40 = 140.0 \ \text{dB}$$

Thus, by Eq. (3.12a), the receiver input power, P_r, is given by

$$P_r = 30 - 3 + 40.5 - 140 - 0.5 + 37 - 2.5$$
$$= -38.5 \ \text{dBm}$$

Received input power P_r resulting from free space loss only is normally designed to be significantly higher than the minimum receivable or threshold level R_{th} for acceptable probability of error. This power difference is engineered so that if the signal fades, due to various atmospheric and terrain effects, it will fall below its threshold level for only a small fraction of time. This built-in level difference is called the link *fade margin*, and thus given by

$$\text{Fade margin} = P_r - R_{th} \qquad (3.13)$$

To give some sense of its magnitude, it is typically designed to be between about 30 and 40 dB, depending on the path reliability desired. When only thermal noise is considered as the degrading factor to bit error rate performance during fading, then the fade margin is referred to as the *thermal fade margin* (TFM). For this situation, R_{th} is usually defined as the signal level, distorted only by thermal noise, that results in a probability of bit error, or *bit error rate (BER)* of 10^{-3}. It should be noted that R^{th} corresponding to a BER of 10^{-3} is a dynamic threshold, used for outage calculations (Sect. 3.8) and "hands-off" field measurements in a normal fading environment. Another threshold used by industry is a static threshold and corresponds to a BER of 10^{-6}. This threshold, which is sometimes referred to as the receiver sensitivity, is measured manually in the factory or in the field in a non-fading environment by inserting an attenuator in the signal path and reducing the received level until the associated BER is achieved.

A useful first-order measure of the performance of a wireless link is its *system gain*. System gain is defined as the difference between the transmitter output power and the receiver threshold level, normally that corresponding to a BER of 10^{-3} or 10^{-6}. Since by Eqs. (3.12b) and (3.13), thermal fade margin can also be expressed as system gain minus net path loss, then, the larger the system gain, the larger the thermal fade margin for a given path. System gain ranges from a high in the region of 130 dB for low modulation order systems operating in the 6 GHz band to a low in the region of 65 dB for high modulation order, high-capacity systems operating in the 80 GHz range.

3.5   Fading Phenomena

3.5.1   Atmospheric Effects

As wireless signals travel along a path from a transmitting to a receiving antenna, they are impacted by atmospheric factors that often distort their transmission relative to true free space propagation. These factors can be grouped into three main categories, namely, refraction, reflection, and absorption. As will be seen, the first impacts transmission primarily below approximately 12 GHz, the second impacts all transmission, and the third impacts transmission primarily above about 12 GHz.

3.5.1.1   Refraction

In ideal free space propagation, it is assumed that a radio signal travels in a straight line. However, a terrestrial radio signal seldom travels in a truly straight line. This is because *atmospheric refraction* causes it to bend slightly from its ideal straight-line path. This bending is due to changes in the atmospheric refractive index with height.

The *refractive index* η of a given medium, which is a function of density, determines the speed, v, of an electromagnetic wave through that medium. Specifically,

$$v = \frac{c}{\eta} \tag{3.14}$$

where

$c =$ speed of light in a vacuum

All electromagnetic waves are bent via refraction when they pass from a medium of one refractive index to another. Figure 3.9 is a simplified depiction of the refraction of a radio wave as it travels from a high density (high refractive index) atmosphere to a low density (low refractive index) atmosphere. As rays at the edge A of the wavefront AB enter the less dense area, they start to travel at a faster speed than rays still traveling in the more dense area. As a result, when the wavefront AB is fully in the less dense medium, the new wavefront $A'B'$ is tilted downwards as AA' is longer than BB'. Most of the time, the refractive index of the atmosphere decreases gradually and uniformly with height as a result of the density of air and its vapor content decreasing gradually and uniformly with height. This thus has the effect of curving a wave in a gradual downward direction as it travels from a transmitter to a receiver. For given antenna heights, this allows the wave to be transmitted further than if it traveled in a straight line by simply tilting the antennas slightly upwards. Figure 3.10 shows this path lengthening effect.

Fig. 3.9 Refraction of a radio wave

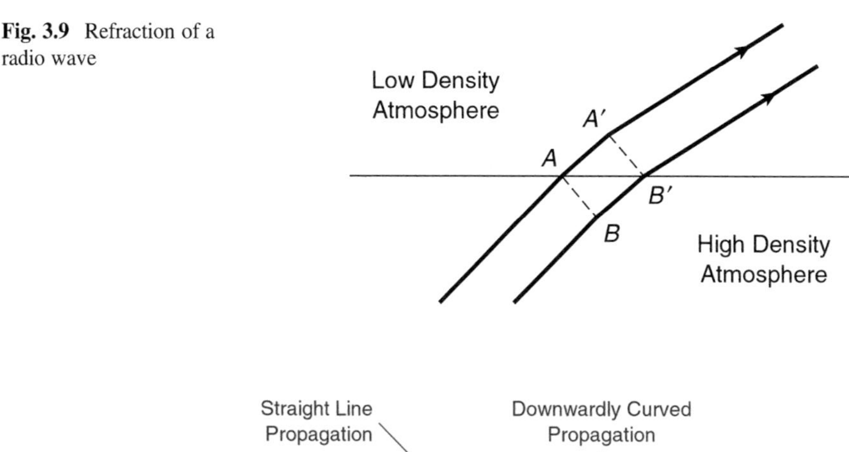

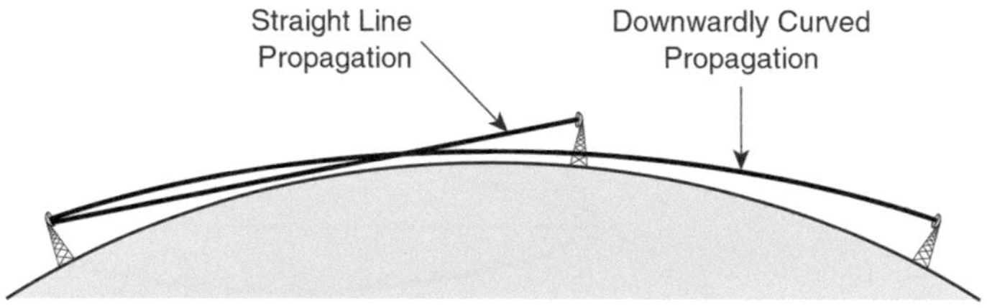

Fig. 3.10 Lengthened path due to refraction-induced downward curving of radio wave

The downward curvature resulting from refraction in a normal atmosphere can be visualized by assuming that the radio wave does in fact travel in a straight line but over a section of the earth's surface that has an effective radius that is greater than that of the true earth. The magnitude of the curving of a transmitted signal over a path is characterized by the path's *earth radius factor, k,* where

$$k = \text{effective earth radius/true earth radius} \tag{3.15}$$

For a value of k of 4/3, for example, a propagation length approximately 15% longer than straight-line propagation is achievable. During normal atmospheric conditions, k ranges from about 1 in dry, elevated areas to 4/3 in typical inland areas, to 2–3 in hot humid coastal areas. However, in abnormal atmospheric conditions, it can vary significantly from its normal value, from a low of about 1/2 to a positive high of infinity and even to where it becomes a negative value. For values of k less than 1, the atmosphere is said to be *substandard* or *subrefractive*, whereas, for values of k greater than 3 or negative, it is said to be *superstandard* or *superrefractive*.

The effects of differing values of k on a radio path are depicted in Fig. 3.11. Any value of k greater than 1 results in a downward curving of the transmitted signal as shown in Fig. 3.11a. From a fictitious straight line radio signal point of view, as k increases above 1 the earth's curvature appears to be flattening, as shown in Fig. 3.11b, until, when k equals infinity, it appears to be totally flat, permitting the radio wave to travel in parallel with it until obstructed. Further downward bending of the wave leads to a depressed earth's surface or negative earth curvature from a straight-line radio signal point of view and hence a negative value of k. When

Fig. 3.11 Effect of differing values of k on radio paths

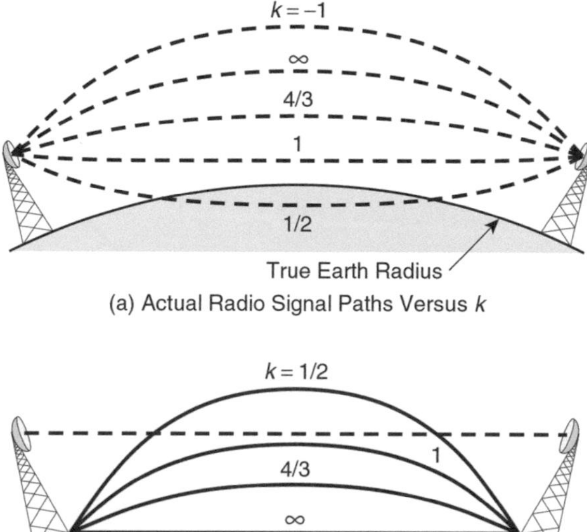

(a) Actual Radio Signal Paths Versus k

(b) Effective Earth Profiles Versus k

k equals 1, no refraction occurs and the radio signal travels in a straight line over the curved earth. For values of *k* less than 1 the transmitted signal curves upward, causing the earth's curvature from a straight-line signal point of view to appear to be bulging, possibly to the extent that it obstructs the path of the radio signal. Values of *k* less than 1 arise when atmospheric density increases with height instead of decreasing, as is normally the case, and can occur when there is heavy fog or extremely cold air over less dense air near the ground, as occurs with a cold front passage.

In addition to the gradual bending of the direct beam described above, refraction can result in additional signal impairment, namely, ducting and multiple refractive paths.

Ducting is an atmospheric phenomenon that can occur when values of *k* go negative. In this state, the atmosphere can act like a horizontal duct or waveguide, which, if located at the general level of the direct radio signal, can trap it within its boundaries. Ducts can be ground-based or elevated.

In ground ducts, the region close to the ground has a high refractive index, which changes rapidly to a low refractive index with height. If a transmitted signal finds itself in such an environment it is bent downwards with a radius smaller than that of the earth. Thus, if the receiving antenna is located beyond the point where the signal hits the earth, a fade large enough to cause a complete outage would likely occur, unless the signal is fortuitously reflected and picked up by the antenna. Figure 3.12a shows a signal trapped in a ground duct with a possible reflected ray. Ground ducts tend to occur most frequently over or near large expanses of water as a result of either evaporation or advection, advection being the movement of one type of air over

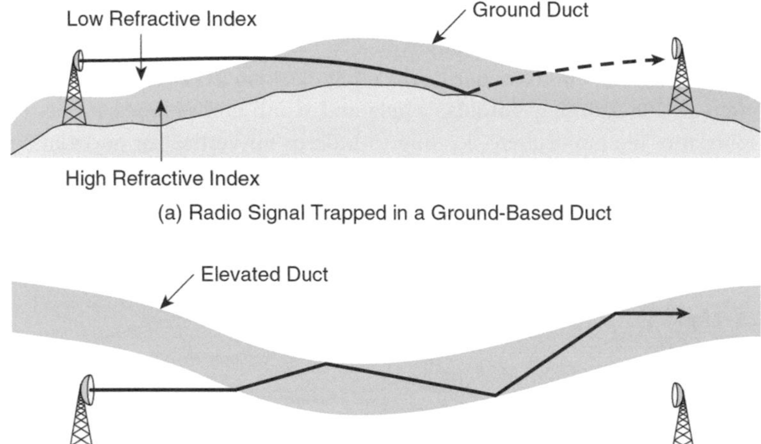

Fig. 3.12   Ducting phenomena

another. Evaporation can lead to an unusually rapid decrease in water vapor density with height resulting in a high-humidity, high-density, high-refractive-index zone below a region of drier, low refractive atmosphere. Such a phenomenon is most likely to occur in the afternoon, as a result of the cumulative effect of solar heating during the preceding period of the day. Ducts formed via convection are typically about 15 m in height. If dry air above land is blown over, moist air above an expanse of water advection occurs, a region of low density, low refractive index being created above a region of high humidity, high refractive index. Such advection usually occurs in the evening when breezes tend to blow from land to sea. Advection ducts are normally about 25 m high.

Abnormal atmospheric conditions, such as a rise in temperature with increasing height, can also result in elevated ducts. Strong elevated ducts, more prevalent in tropical climates, can trap radio signals within its boundaries over relatively long distances, sometimes far beyond the horizon. Figure 3.12b shows how a signal can be trapped in such a duct. Once within the duct, it's reflected off the duct's boundaries and propagates within the duct much like how light travels in a fiber optic glass cable.

Of all the atmospheric anomalies that impact digital wireless transmission performance, none, in general, has greater impact on reliability than those that result in *multiple refractive paths* between the transmitter and receiver. These anomalies occur when there is minimal normal atmospheric turbulence, thus permitting the formation of non-uniform vertical distributions of temperature and humidity. This in turn leads to significant variability with height, or, as it's sometimes referred to, stratification, in the refractive index vertical profile. This situation permits energy, leaving the radiating antenna at different angles, to travel to the receiving antenna via slightly different paths, each with a different length and hence different time delay. This stratification can be considered as a mild form of ducting. The creation of multiple refractive atmospheric paths typically begins to occur on clear, calm, hot summer evenings and builds up during the night, peaking in the early morning hours. Thereafter, as the morning unfolds, winds and rising convection currents tend to thoroughly mix the atmosphere, leading to little or no vertical or horizontal variations in the refractive index gradient in the vicinity of a typical path. As a result, only a single dominant path then tends to exist between transmitter and receiver. Figure 3.13 shows a radio link with multiple refractive atmospheric paths.

Fig. 3.13 Link with multiple refractive atmospheric paths

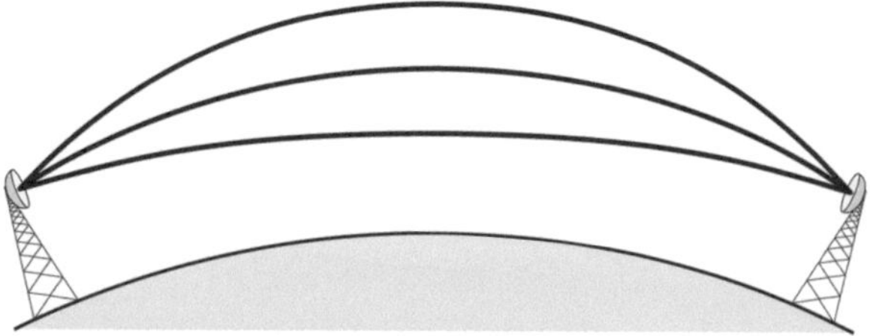

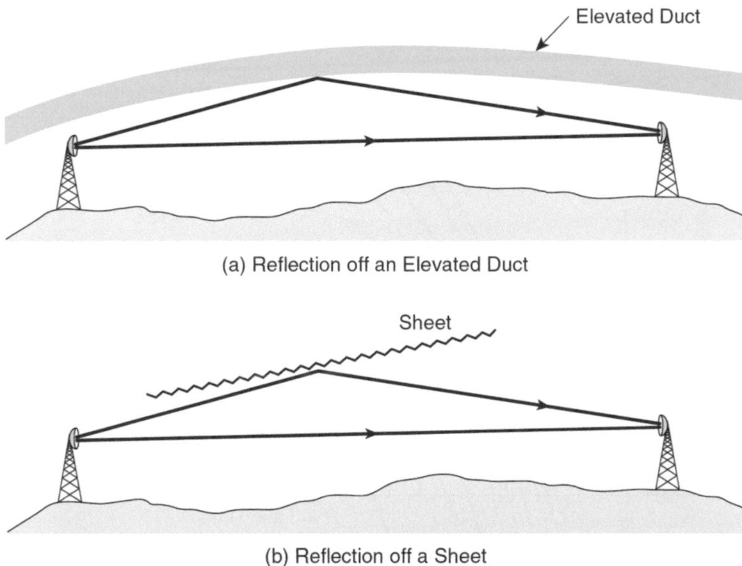

Fig. 3.14 Atmospheric reflections

3.5.1.2   Reflection

In addition to traversing down their direct path, radio signals can reach the receiving antenna via *atmospheric reflection*. Signals can be reflected off ground ducts and elevated ducts described above. The boundaries of these ducts are usually relatively stable, normally moving slowly in an up and down direction. Signals can also be reflected off *sheets*. Sheets are high-altitude undulating layers that are several miles long and constantly changing, often in a rapid fashion, in height and location. Figure 3.14a shows a path with signals reflected off an elevated duct, and Fig. 3.14b shows a path with a signal reflected off a sheet.

3.5.1.3   Rain Attenuation and Atmospheric Absorption

As a radio signal propagates down its path, it may find itself subjected, in addition to the possible effects of refraction and reflection, to the attenuating effects of rain, sleet, snow, and hail and absorption by atmospheric gasses, primarily water vapor and oxygen.

Raindrops attenuate a radio signal by absorbing and scattering the radio energy, with this effect becoming more and more significant as the wavelength of the signal decreases towards the size of the raindrops. Figure 3.15, which is from Rogers et al. [2], shows attenuation in dB/km versus frequency for several rain conditions. As can be seen from the figure, attenuation increases with propagation frequency and rain intensity. At 6 GHz, the attenuation due to a heavy cloudburst is about 1 dB/km. For

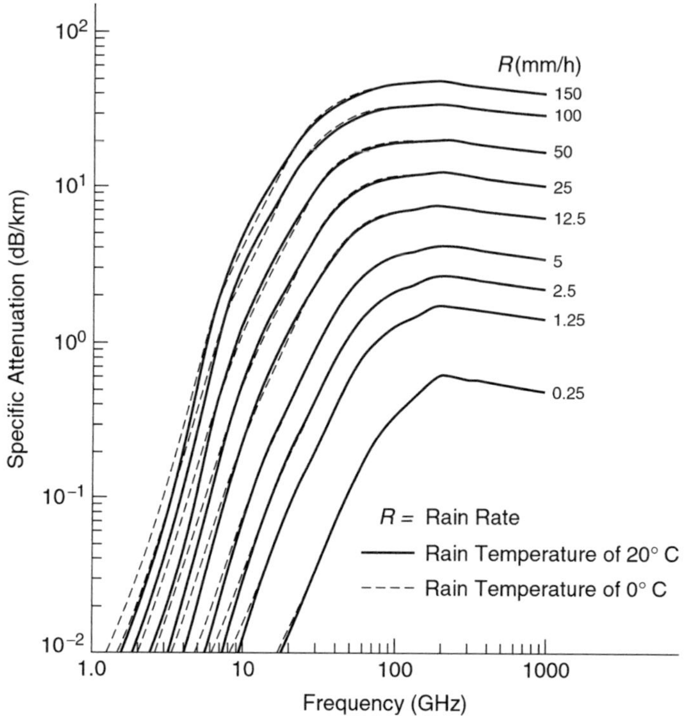

Fig. 3.15 Attenuation due to rain. (From [2] with the permission of the Communications Research Center, Canada)

a 40 km path, this would imply an attenuation of about 40 dB if the cloudburst extended across the entire path. Fortunately, the width of cloudburst cells tends to be on the order of kilometers, so maximum attenuation due to rain at these frequencies tends to be less than 10 dB, an amount easily accommodated by a built-in fade margin. In general, rain attenuation is not a major problem below about 10 GHz. However, this situation changes measurably as the frequency increases, and normally must be factored in at frequencies above about 10 GHz. For example, at 23 GHz, maximum attenuation due to rain is in the order of 20 dB/km. Thus, for a 23 GHz link, with a 40 dB fade margin and operating in a region where heavy cloudbursts occur, path length may have to be limited to no more than a couple of kilometers, if high reliability is to be maintained. The situation is even more dire at 80 GHz where the maximum attenuation is in the order of 50 dB per kilometer. The results in Fig. 3.15 are theoretically derived for spherical raindrops. Real raindrops, however, are somewhat flattened as they fall through the atmosphere. As a result, they have a smaller size in the vertical plane than in the horizontal one. This in turn leads to the attenuation of a vertically polarized wave being somewhat less than that of a horizontally polarized one. A more precise estimate of rain attenuation that factors in the polarization of the signal is given in [3].

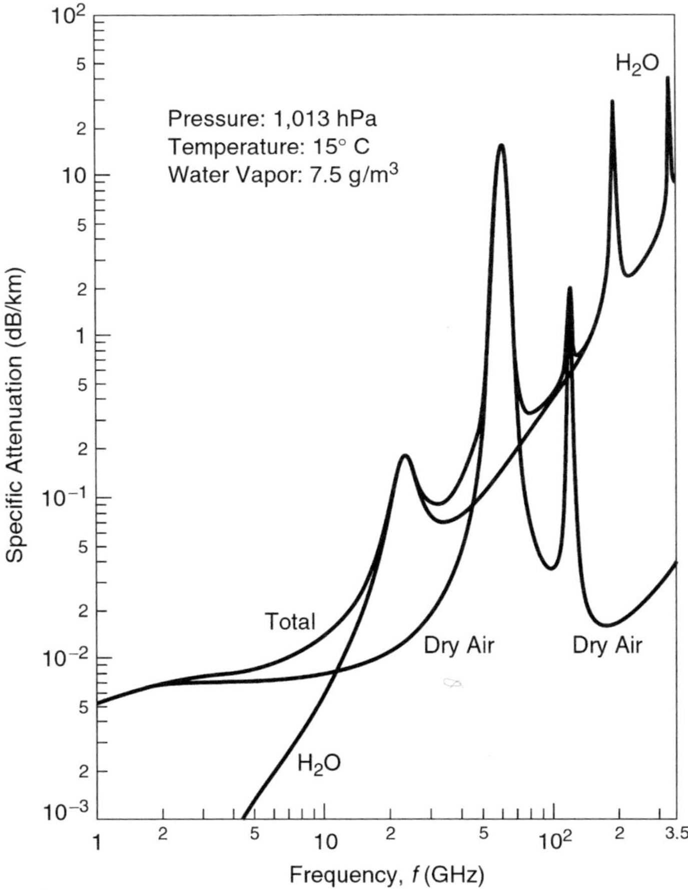

Fig. 3.16 Specific attenuation due to atmospheric gasses. (From [4] with the permission of the ITU)

Figure 3.16, which is from ITU Recommendation ITU-R P.676-12 [4], shows attenuation in dB/km due to water vapor and dry air as a function of frequency in a low to moderate humidity region. For wireless transport systems, the highest frequencies currently in use are in the 80 GHz range. From the figure, it will be observed that attenuation due to water vapor has a first peak at approximately 22 GHz with a value of about 0.18 dB/km. It then declines with increasing frequency up to about 31 GHz, after which it increases back to about 0.18 dB/km in the region of 60 GHz. Attenuation due to dry air is below 0.01 dB/km for frequencies up to 20 GHz, then increases somewhat exponentially to a peak of approximately 15 dB/km in the region of 60 GHz. Paths at frequencies close to 22 GHz seldom exceed about 15 km in length due to rain attenuation and limited transmitter output power. Thus, the maximum likely loss due to atmospheric absorption will be from water vapor and will be on the order of 3 dB, measurable, but not significant relative to likely rain loss. Paths close to 60 GHz, unfortunately, have the worst of two worlds. Rain attenuation can approach 40 dB/km and is additive to dry air attenuation of approximately 15 dB/km. As a result, fixed wireless systems operating at frequencies in this region and in locations subject to any appreciable rainfall are usually limited in path length to about a kilometer or less.

3.5.2   Terrain Effects

The propagation of radio signals through the atmosphere is affected by the terrain in or close to its path in addition to the atmospheric effects discussed above. Potential propagation affecting terrain features are trees, hills; sharp points of projection; reflection surfaces such as ponds, lakes, and seas; and man-made structures such as buildings and towers. These features can result in either reflected or diffracted signals arriving at the receiving antenna.

3.5.2.1   Terrain Reflection

Reflected signals arriving at a receiving antenna are combined with the direct signal to form a composite signal that, depending on the strengths and phases of the reflected signals, can significantly degrade the desired signal. Figure 3.17 shows a radio path with a direct signal and a reflected signal off a lake. The strength of a reflected signal at a receiving antenna depends on the directivity of both the transmitting and receiving antennas, the heights of these antennas above the ground, the length of the path, and the strength of the reflected signal relative to the incident signal at the point of reflection. The phase of a reflected signal at a receiving antenna is a function of the reflected signal path length and the phase shift, if any, incurred at the point of reflection.

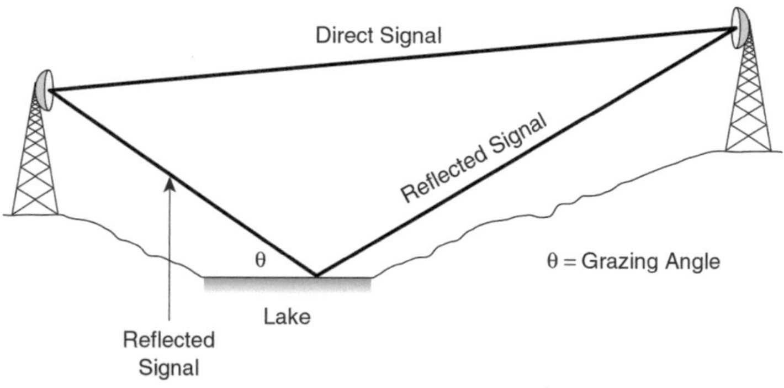

Fig. 3.17 Ground reflected signal

The strength and phase of a reflected signal relative to the incident signal at the point of reflection is strongly influenced by the composition of the reflecting surface, the curvature of the surface, the amount of surface roughness, the incident angle, and the radio signal frequency and polarization. If reflections occur on rough surfaces, they usually do not create a problem as the incident and reflected angles are quite random. Reflections from a relatively smooth surface, however, depending on the location of the surface, can result in the reflected signal being intercepted by the receiving antenna. The transfer function of a reflective surface is referred to as its *reflection coefficient, R*. For a highly reflective surface (a specular reflection plane such as over water) vertically polarized signals show more frequency dependence than horizontally polarized ones. However, at frequencies between about 2 and 10 GHz, variation with frequency for both polarizations is sufficiently small that the following general observations can be made:

For horizontally polarized signals, the magnitude of the reflection coefficient is unity
  for grazing angles between zero and a few degrees and then slowly decreases to
  about 0.8 at the maximum grazing angle of $90°$. Thus, for practical fixed wireless
  links, where grazing angle rarely, if ever, exceeds a couple of degrees, the
  magnitude of the reflection coefficient can be treated as unity.
For vertically polarized signals, the magnitude of the reflection coefficient is unity
  for zero grazing angle, but falls off rapidly, to a value of about 0.9 at $0.4°$ grazing
  angle, and on to a minimum in the region of 0.1 at a grazing angle of about 6 to $7°$.
  It then increases back up to about 0.8 at a grazing angle of $90°$.

On overwater paths at frequencies above about 3 GHz, it is advantageous to choose vertical over horizontal polarization. With vertical polarization and grazing angles greater than about $0.7°$, a reduction in the surface reflection of 2–17 dB over that with horizontal polarization can be expected [5].

A method to calculate effective surface reflection coefficient is given in [5] by the ITU.

3.5.2.2   Fresnel Zones

The proximity of terrain features to the direct path of a radio signal impacts their effect on the composite received signal. *Fresnel zones* are a way of defining such proximity in a very meaningful way. In the study of the diffraction of radio signals, the concept of Fresnel zones is very helpful. Thus, before reviewing diffraction, an overview of Fresnel zones is in order. The first Fresnel zone is defined as that region containing all points from which a wave could be reflected such that the total length of the two-segment reflected path exceeds that of the direct path by half a wavelength, $\lambda/2$, or less. The nth Fresnel zone is defined as that region containing all points from which a wave could be reflected such that that the length of the two-segment reflected path exceeds that of the direct path by more than $(n-1)\lambda/2$ but less than $n\lambda/2$. The boundary of a Fresnel zone surrounding the direct path in the plane of the path is an ellipsoid whereas it is a circle in the plane perpendicular to the

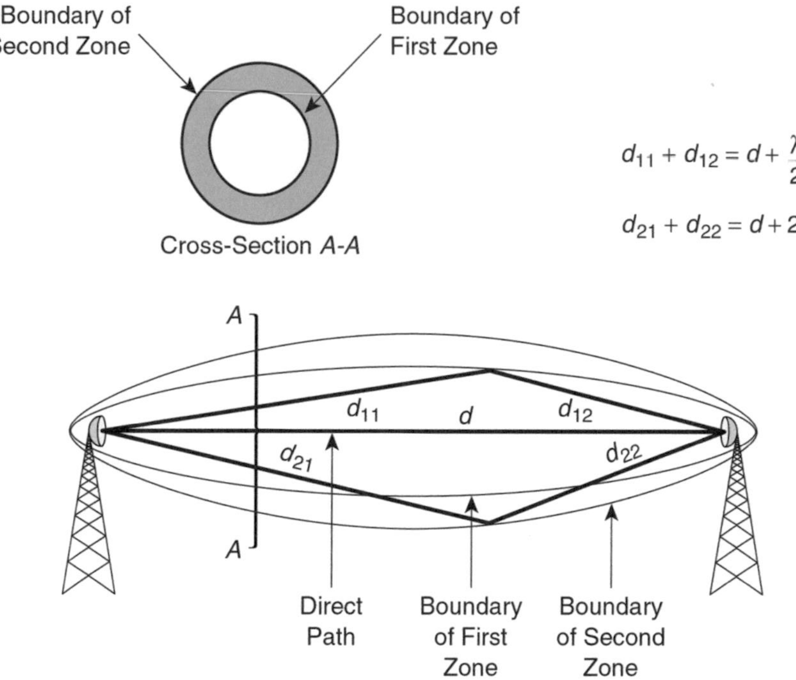

$$d_{11} + d_{12} = d + \frac{\lambda}{2}$$

$$d_{21} + d_{22} = d + 2 \cdot \frac{\lambda}{2}$$

Fig. 3.18 First and second Fresnel zone boundaries

path. Figure 3.18 shows the first and second Fresnel zone boundaries on a line-of-sight path. The perpendicular distance F_n from the direct path to the outer boundary of the nth Fresnel zone is approximated by the following equation:

$$F_n = \left[\frac{n\lambda d_{n1} d_{n2}}{d}\right]^{\frac{1}{2}} \tag{3.16}$$

where

d_{n1} = Distance from one end of the path to point where F_n is being determined
d_{n2} = Distance from other end of the path to point where F_n is being determined
d = Length of the path = $d_{n1} + d_{n2}$.

and where λ, d_{n1}, d_{n2} and d are measured in identical units.

Specifically, F_1 is given by

$$F_1 = 17.3 \left[\frac{d_{11} d_{12}}{fd}\right]^{\frac{1}{2}} \text{ m} \tag{3.17a}$$

where d_{11}, d_{12}, and d are in km and f is in GHz.

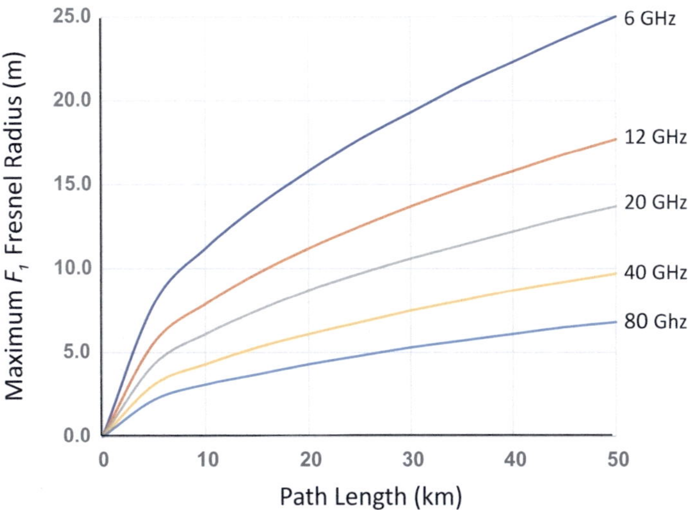

Fig. 3.19 Maximum value of F_1 versus distance and frequency

And given by

$$F_1 = 72.1 \left[\frac{d_{11}d_{12}}{fd} \right]^{\frac{1}{2}} \text{ ft} \tag{3.17b}$$

where d_{11}, d_{12}, and d are in miles and f is in GHz.

Figure 3.19 shows the maximum value of F_1 versus path length and frequency. We note from this figure that for a 50 km path operating at 6 GHz, the maximum value of F_1, which occurs at the middle of the path, is 25 m, whereas for a 5 km path operating at 40 GHz, F_1 maximum is only 3 m.

Most of the power that reaches the receiver is contained within the boundary of the first Fresnel zone (more on this in the next section). Thus, terrain features that lie outside this boundary, with the exception of highly reflective surfaces, do not, in general, significantly affect the level of the received signal. For a signal reflected at this boundary that acquires a $\lambda/2$ phase shift as a result of the reflection, its total phase shift at the receiving antenna relative to the direct signal is λ, and hence, the reflected signal is additive to the direct signal. However, for a signal that experiences zero phase shift, as is possible with a vertically polarized reflected signal with a large incident angle, the total relative phase shift at the receiving antenna is $\lambda/2$ and hence results in a partial cancellation of the direct signal. The maximum increase possible in signal strength resulting from a reflected signal is 6 dB. However, the maximum loss possible is in theory an infinite number of decibels, and in practice can exceed 40 dB.

3.5.2.3 Diffraction

So far in explaining propagation effects it has been tacitly assumed that the energy received by a radio antenna travels as a beam from transmitting to receiving antenna that's just wide enough to illuminate the receiving antenna. This is not exactly the case, however, as waves propagate following *Huygen's principle*. Huygen showed that propagation occurs along a wavefront, with each point on the wavefront acting as a source of a secondary wavefront known as a wavelet, with a new wavefront being created from the combination of the contributions of all the wavelets on the preceding front. Importantly, secondary wavelets radiate in all directions. However, they radiate strongest in the direction of the wavefront propagation and less and less as the angle of radiation relative to the direction of the wavefront propagation increases until the level of radiation is zero in the reverse direction of wavefront propagation. The net result is that, as the wavefront moves forward it spreads out, albeit with less and less energy on a given point as that point is removed further and further from the direct line of propagation. This sounds like bad news. The signal intercepted at the receiving antenna, however, is the sum of all signals directed at it from all the wavelets created on all wavefronts as the wave moves from transmitter to receiver. At the receiver, signal energy from some wavelets tends to cancel signal energy from others depending on the phase differences of the received signals, these differences being generated as a result of the different path lengths. Likewise, signal energy from some wavelets tend to be additive with signal energy from others. Signal components that have a phase difference of $\lambda/4$ or less relative to the direct line signal are additive. Signals with a phase difference between $\lambda/4$ and $\lambda/2$ are subtractive. In an unobstructed environment, all such signals fall within the first Fresnel zone. In fact, the first Fresnel contains most of the power that reaches the receiver. Consider now what happens when an obstacle exists in a radio path within the first Fresnel zone. Clearly, under this condition, the amount of energy intercepted at the receiving antenna will differ from that intercepted if no obstacle were present. The cause of this difference, which is the disruption of the wavefront at the obstruction, is called *diffraction*. If an obstruction is raised in front of the wave so that a direct path is just maintained, the power reaching the receiver will be reduced, whereas the simplistic narrow beam model would suggest that full received signal would be maintained. A positive aspect of diffraction is that if the obstruction is further raised, so that it blocks the direct path, signal will still be intercepted at the receiving antenna, albeit at a lesser and lesser level as the height of the obstruction increases further and further. Under the simplistic narrow beam model, one would have expected complete signal loss. Figure 3.20a shows "unobstructed" free space propagation, where path clearance is assumed to exceed several Fresnel zones. Figure 3.20b shows diffraction around an obstacle assumed to be within the region of the first Fresnel zone but not blocking the direct path. Figure 3.20c shows diffraction around an obstacle blocking the direct path. For simplicity, an expanding wave that has progressed partially down the path is shown in all three depictions, to the point of obstruction in the case of the obstructed paths. All wavelets on the

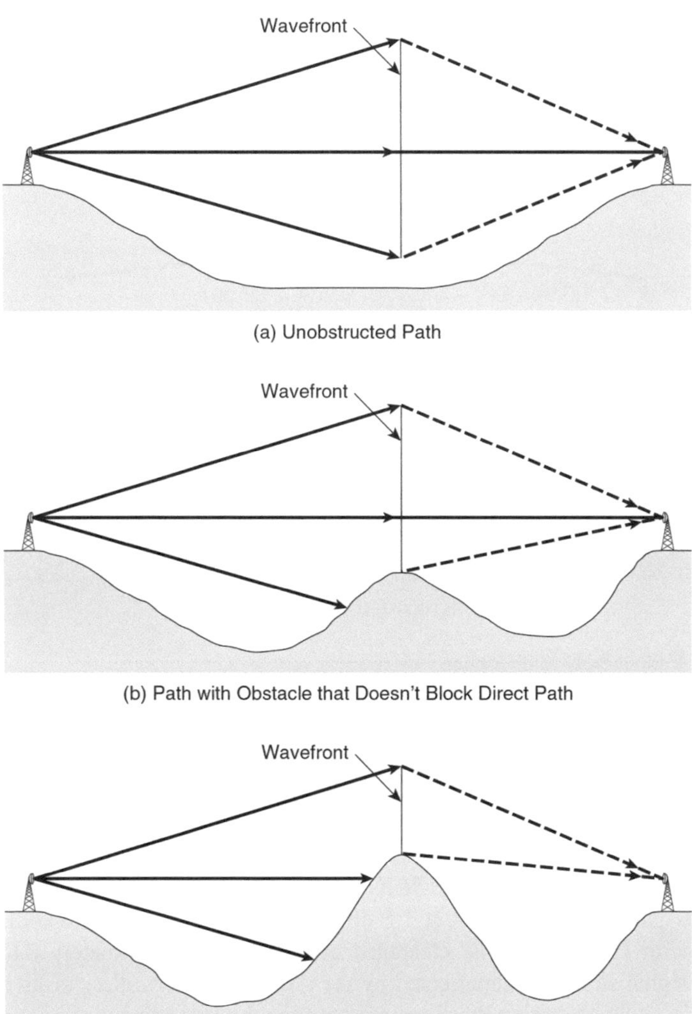

Fig. 3.20 Graphical representation of diffraction

wavefront shown in the unobstructed case are able to radiate a signal that falls on the receiving antenna. In the case of the obstructed paths, however, the size of the wavefronts and hence the number of wavelets radiating signals that fall on the receiving antenna are decreased.

For analytical purposes, diffracting terrain is normally classified as being one of three types, namely, smooth (spherical) earth, rounded obstacle and knife-edge. In reality, however, terrain property normally falls somewhere between these hypothetical types. Figure 3.21 shows, for each of these types of diffracting terrain, an example of diffraction around such terrain that's blocking a path. Like loss due to

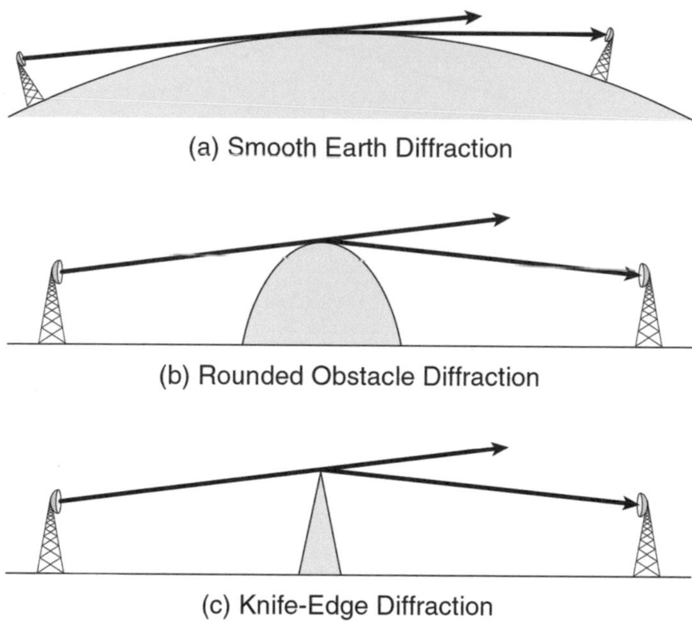

Fig. 3.21   Various types of diffraction

reflection, diffraction loss is a function of the nature of the diffracting terrain, varying from a minimum for a single knife-edge obstacle to a maximum for smooth earth. Much research and analysis of diffraction loss has been conducted over several decades leading to well accepted formulae [6] for its estimation. For the situation where the clearance between an obstacle and the direct path is equal to F_1, the first Fresnel zone clearance at the obstacle location, then there is, in fact, a received signal gain. This gain varies from about 1.5 dB for the knife edge case to 6 dB for the smooth earth case. When the clearance decreases to approximately $0.6\, F_1$, the received signal strength is unaffected by the obstruction, regardless of its type. For the grazing situation, where the clearance between the obstacle and the direct path is reduced to zero, diffraction loss varies from approximately 6 dB for the knife-edge case to approximately 15 dB for the smooth earth case.

3.5.2.4   Path Clearance Criteria

Radio paths must be designed to minimize as much as possible the likelihood of significant loss in received signal level due to obstructions in the path. In so designing, the designer must take into account the fact that the effective clearance, F, between the obstruction and the direct path varies as the earth radius factor, k, varies. For a path where the normal value of k is 4/3, for example, then whenever k changes to a value less than 4/3, F diminishes, resulting in increased obstruction

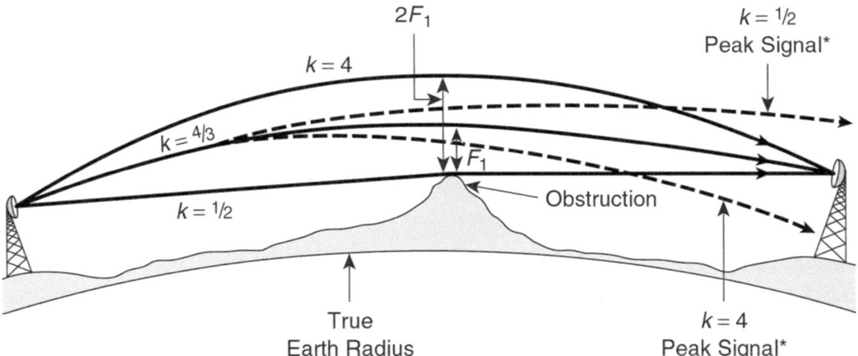

Fig. 3.22 Impact of varying k on clearance

loss if the obstruction lies within the first Fresnel zone. Figure 3.22 shows the impact of varying k on clearance F. In this example, for $k = 4/3$, F is equal to F_1. When k decreases to 1/2, F reduced to zero, whereas when k increases to 4, F doubles to $2F_1$. Note, also, that if the antennas are adjusted for maximum signal strength when k equals 4/3, then when k equals 1/2, the peak transmitted signal passes above the receiving antenna and a slightly weaker direct path signal (a function of the transmitting antenna directivity) arrives at the receiving antenna. The receiving antenna further attenuates this signal as it's positioned to provide maximum gain to the k equals 4/3 signal. When k is 4, the peak direct signal now passes below the antenna and again the antennas will not be optimally aligned and thus a slight reduction in direct path signal strength will result.

A number of planning procedures have been developed in different countries and by different regulating agencies in terms of the clearance of the controlling obstruction, relative to F_1, so as to guarantee, to a large extent, that diffracted signals do not result in a significant signal loss under most atmospheric conditions. Among such procedures is one developed by the ITU and published in its recommendation ITU-R P.530-17 [5]. As the cost of a radio tower tends to increase significantly with height, and as high towers stand out environmentally, the radio path designer typically chooses antenna heights that result in these or similar criteria just being met or being met with only a small margin.

3.5.3   Signal Strength Versus Frequency Effects

Fading is the variation in strength of a received radio signal due to the atmospheric and terrain effects in the radio's path as discussed in Sections 3.5.1 and 3.5.2. Fading is normally broken down into two main types, namely, *flat fading* and *frequency selective fading*. In flat fading, the signal is attenuated uniformly across the frequency band occupied by the signal, whereas in frequency selective fading, the attenuation varies with frequency across the occupied band. These two types of fading can occur separately or together, though the latter is rare. Exactly when fading resulting from variations in atmospheric conditions is likely to occur cannot be predicted with any accuracy. Such fading, however, can be predicted on a statistical basis, given a statistical knowledge of the atmospheric conditions in the vicinity of the radio path.

3.5.3.1   Flat Fading

Beam Bending Fading   The fading resulting from beam bending which in turn results from slow changes in the K factor is flat fading. As discussed previously, the impact of such possible fading is normally minimized by choosing antenna heights that result in the meeting or exceeding of path clearance criteria.

Duct Entrapment Fading   Fading as a result of duct entrapment is also flat fading. Unfortunately, other than keeping the path length short to minimize the probability of entrapment, no effective countermeasures exist.

Rain Fading   Finally, fading caused by rain is also flat fading. Effective countermeasures are a high fade margin, a path length short enough to guarantee a low probability of the maximum rain rate resulting in path attenuation that exceeds the fade margin, and the use of vertical polarization on paths operating above 10 GHz, if possible. As indicated in Sec. 3.5.1.3, rain attenuation to vertically polarized signals is less than that to horizontally polarized ones. Thus, choosing vertical polarization reduces rain-induced outages (by from 40 to 60%), compared to horizontal polarization.

3.5.3.2   Frequency Selective Fading

Ground Reflection Fading   If a radio path passes over highly reflective ground or water, and the antenna heights allow a reflected path between the antennas, reflection fading is likely and, under certain circumstances, can be substantial. Ground reflection fading is frequency selective. The reflected signal will have, at any instant, a path length difference and hence a time delay, τ say, relative to the direct path that's independent of frequency. However, this delay, measured as a phase angle, varies with frequency. Thus, if the propagated signal occupies a band between frequencies f_1 and f_2 say, then the relative delay in radians at frequency f_1 will be $2\pi f_1 \tau$, the relative delay at f_2 will be $2\pi f_2 \tau$, and the difference in relative phase delay between the component of the reflected signal at f_1 and that at f_2 will be $2\pi(f_2 - f_1)\tau$. Because the reflected signal will have a phase shift relative to the direct signal that's a function of frequency, then when both these signals are combined a different resultant signal relative to the direct signal will be created for each frequency. Depending on the value of τ and the signal bandwidth $f_2 - f_1$, significant differences can exist in the composite received signal amplitude and phase as a function of frequency as compared to the undistorted direct signal. Further, the closer the reflected signal is in amplitude to the direct signal, the larger the maximum signal cancellation possible, this occurring when the signals are $180°$ out of phase.

Normally, on a path with a highly reflective surface, the path designer will first try to shift the reflection point from where a reflected signal can be intercepted by a receiving antenna away from the highly reflective surface. This is done by adjusting the relative heights of the antennas. Alternatively, if an appropriate obstruction exists, the designer may locate the height of one antenna so that the obstruction denies the reflected signal a line-of-sight to the receiving antenna. Highly directive antennas may help somewhat as they reduce the strength of the received reflected signal relative to that resulting from less directive antennas. Improvement due to antenna discrimination may also be achieved by tilting one or both antennas slightly upwards. However, in so doing the strength of the direct signal is slightly decreased. For overwater paths with an unblocked reflected signal, if path geometry is such that the grazing angle is greater than about $0.7°$, then the use of vertical polarization can significantly reduce the maximum possible depth of reflection-induced fading as compared to the horizontally polarized case. This is because, as pointed out in Sect. 3.5.2.1, the amplitude of the reflection coefficient for vertically polarized signals falls off rapidly beyond this grazing angle, reducing the cancellation effect at the receiver. For a horizontally polarized signal, however, the magnitude of the reflection coefficient remains very close to unity, resulting in the possibility of significant signal cancellation. As indicated earlier, ITU Recommendation ITU-R P.530-17 [5] indicates that, for overwater paths at frequencies above about 3 GHz, it is advantageous to use vertical polarization, noting that "At grazing angles greater than about $0.7°$, a reduction in the surface reflection of 2–17 dB can be expected over that at horizontal polarization."

If the implementation of the above techniques, where possible, does not result in acceptable path performance, then the normal method used to improve reliability is space diversity reception (Sect. 3.8.1).

Atmospheric Multipath (Rayleigh) Fading Atmospheric refractive multipaths are the dominant factor in fading for paths operating at about 10 GHz or below, assuming adequate clearance has been provided to minimize beam bending and obstruction-related fading. This being so, going forward in this text, unless indicated otherwise, the term multipath fading will be used to mean fading resulting from atmospheric multipaths. For the same reason given above for ground reflection fading, multipath fading is frequency selective. To gain a first-order insight into multipath fading, let us consider a simple two-ray model, with a direct ray and a refracted ray. If τ is the relative time delay between the direct and the refracted ray, then the relative phase delay between these rays differs by $2\pi(f_2 - f_1)\tau$ between f_1 and f_2, the signal band edge frequencies. The amplitude response versus frequency of this two-ray transfer function will be repetitive over a frequency difference $\Delta f = 1/\tau$, since this results in a phase change of 2π radians. If the bandwidth of the signal is very much less than $1/\tau$, then the amplitude and phase variation across it will be

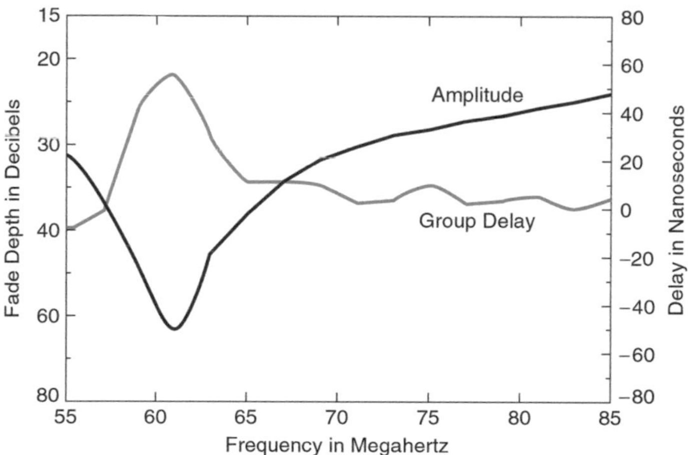

Fig. 3.23 A typical scan of a multipath fading event in a 6 GHz radio channel. (From [7], with the permission of the IEEE)

small and the resulting multipath fading can be treated as flat fading. As Δf approaches about $1/10\tau$, however, the frequency selective effects become significant and cannot be ignored. Figure 3.23, which is from Rummler et al. [7], shows a typical scan of a multipath fading event on a typical 30 MHz bandwidth, 6 GHz radio channel. It will be observed that the fade depth varies from about 25 dB to about 65 dB within the band. The group delay, which is the rate of change of phase, and is constant with frequency for a linear non-distorting transfer function, similarly shows significant variation across the band. Inband fade depth variation similar to that shown here is quite common on paths subject to heavy multipath fading.

Multipath fades can be quite deep, often exceeding 40 dB. It is shown in Fig. 4.11 of [8] that for a 6 GHz path in Georgia, United States, these fades take place quite fast, typically on a scale of seconds, and occur quite randomly. Fortunately, fade duration is inversely proportional to fade depth. For example, on a 4 GHz system with a path length of 30–35 miles, the median duration of 20 dB fades is about 30 s, whereas the median duration of 40 dB fades is about 3 s [9]. Figure 4.12 of [8] shows that on the same path from which the data in Fig. 4.11 of [8] was derived, for fade depth deeper than about 20 dB, the fade time decreases by a factor of about 10 for a 10 dB increase in fade depth. Such a distribution of received signal power versus time the power is exceeded is a Rayleigh distribution [10]. As a result, multipath fading is often referred to as Rayleigh fading. Measurements of multipath fading indicate that the number of discrete fades, as well as the percentage of time a fade exceeds a given level, tends to increase with path length and frequency. During periods of heavy rainfall, the right atmospheric conditions do not normally exist for multipath fading. Thus, at any given time, if fading is present, it is usually multipath or flat, but not both simultaneously.

3.5.3.3   Multipath Fading Channel Model

In this section, a highly accepted model for simulating the frequency selective effects of multipath propagation is reviewed. Such models are very useful in evaluating the performance of digital radio links. Depending on their bandwidth and modulation complexity, such links can be highly sensitive to the frequency selective distortion resulting from multipath fading. This is because the severe amplitude and group delay distortion resulting from multipath fading causes severe intersymbol interference [11]. This in turn results in degraded bit error rate performance beyond that which would be generated by an equivalent flat fade [12]. It is important to recognize that the purpose of defining a channel model is only to describe the frequency response of the channel, not the physical propagation characteristics of the channel. This allows the channel model to be considerably less complex than the actual propagation channel. With a multipath fading model, it becomes possible to estimate the fraction of time, i.e., the probability, that the propagation conditions on the path will be such that the radio link cannot meet a defined minimum acceptable performance, normally defined as a bit error rate (BER) of 10^{-3}.

Many multipath fading channel models have been proposed. However, over time, one model, the simplified three-ray (three-path) model, initially proposed by Rummler [13], has found considerable favor, and this is the only model that will be reviewed here. For an excellent overview of multipath fading models, the reader is referred to Rummler et al. [7], which has provided the basis of this review.

The simplified three-ray model defines the complex voltage transfer function H ($j\omega$) of a multipath channel by the relationship

$$H(j\omega) = a\left[1 - be^{-j(\omega-\omega_0)\tau}\right] \tag{3.18}$$

where

a is a flat loss term and
b is a term related to the notch depth relative to a.

The quantity in brackets suggests interference between two rays with relative delay τ that produces a minimum in the response (a notch) at ω_0, with both ω_0 and ω being measured from a common reference, usually the channel midfrequency.

The flat fade power loss term in decibels, A, is given by

$$A = -20 \log_{10} a \tag{3.19}$$

and the relative notch depth in decibels, B, is given by

$$B = -20 \log_{10}(1 - b) \tag{3.20}$$

Thus, $A + B$ equals the total fade depth at the response minimum which is the center of the notch. Figure 3.24, which is from Rummler et al. [7], shows the amplitude response of $H(j\omega)$ for $\tau = 6.3$ ns.

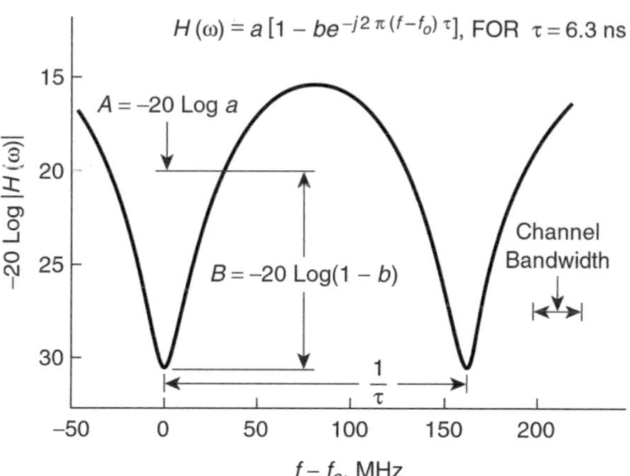

Fig. 3.24 Attenuation of the modeling function used in the simplified three-ray model. (From [7], with the permission of the IEEE)

When the amplitude b has a range from 0 to 1 and τ is positive, the response is said to be *minimum phase*. When b continues to have a range from 0 to 1, but τ is negative, the response is said to be *non-minimum phase*. When b is greater than 1 and τ is positive, the response is also said to be non-minimum phase.

At first glance, Eq. (3.18) suggests a two-ray model, the signal from one path being a and the signal from the other being $abe^{-j(\omega-\omega_0)\tau}$. In fact, this model is sometimes called the two-ray model with flat attenuation. However, this response can also be visualized as a three-ray model. In this visualization, there is a direct ray, ray 1 say, which is unfaded; a second ray, ray 2 say, of similar but varying negative and positive amplitude and close enough in delay to ray 1 that their composite amplitude response, of value a relative to that of ray 1, is flat over the channel width (no frequency selective fading), and a third ray, ray 3, at a relative delay τ to rays 1 and 2 that provides the frequency shaping of $H(j\omega)$. Figure 3.25 depicts the simplified three-ray model, recognizing that this illustration does not depict the physical path.

The strong acceptance of the simplified three-ray model is in part due to the fact, in addition to its simplicity, it has been found to provide a very good fit to almost all measured responses in standard digital radio channels. When this model is being used, the delay parameter τ is usually fixed at a value that results in the period of $H(j\omega)$ (which equals $1/\tau$) being large compared to the channel bandwidth under consideration. In the original development of this model, τ was chosen so that the period of $H(j\omega)$ was six times the channel bandwidth under consideration. This led to a value of τ of 6.3 ns. This value has now been accepted as "standard" by many, but some still apply the factor of six rule which gives a value of τ that's related to the specific channel bandwidth under consideration. Note that since the model is not representative of the physical path, but purely a mathematical way to predict the path transfer function $H(j\omega)$, no physical interpretation should be placed on τ.

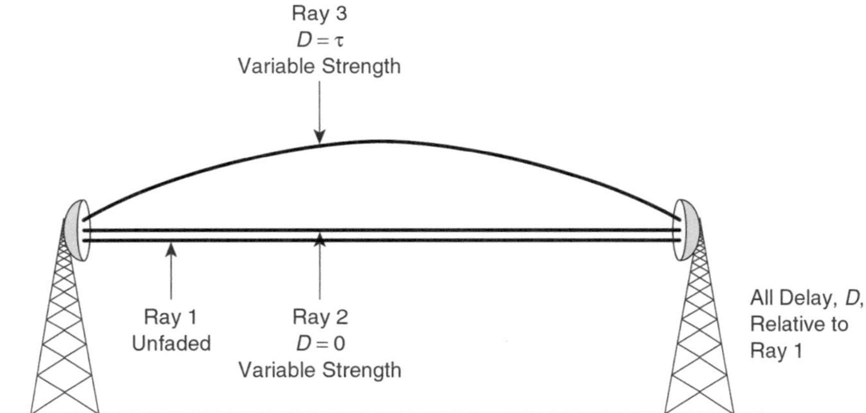

Note: This model is not representative of the physical path. It is a model to predict
the path transfer function H(ω) only.

Fig. 3.25 Depiction of the simplified three-ray model

In order to be able to calculate the outage time of a specific radio link due to
multipath fading, one must find a way to characterize the radio performance in the
presence of such fading. The channel model provides a vehicle for this characterization, the radio performance being defined in terms of the model function parameters. Since the frequency selective effect of multipath fading on a signal is often
referred to as dispersion, such a characterization is called the radio's *dispersion
signature*, and is defined for a specific BER performance threshold, typically either
10^{-3} or 10^{-6}. Applying the simplified three-ray model, the radio is characterized in
the laboratory via a path simulator that emulates Eq. (3.18) with $a = 1$ (achieved by a
high fixed direct signal level which minimizes the effect of thermal noise), while
varying b and ω_0 to obtain a BER of either 10^{-3} or 10^{-6}. Thus, the signature is a
graph of fade notch depths B for a BER of 10^{-3} or 10^{-6} versus the frequency from
the channel center.

A radio receiver BER performance in the presence of multipath fading differs
depending on whether the propagation condition is a minimum phase or non-
minimum phase. Figure 3.26, which is from Rummler et al. [7], shows the minimum
phase and non-minimum phase dispersion signature of a 16-QAM radio for a BER of
10^{-3}. For all points below the signature, the radio BER is higher than 10^{-3}. It will be
noted that the minimum phase signature is narrower and lower than that for
non-minimum phase. It thus requires greater notch depths, particularly at the band
edges, to be degraded to a BER of 10^{-3} compared to the non-minimum phase case.
Thus, for the radio characterized here, minimum phase performance is clearly
superior to non-minimum phase. This is normally the case in the absence of effective
channel equalization (Sect. 5.9).

For outage calculations, the signature is characterized in terms of signature width
and signature depth. Here, signature width is defined as the distance between the

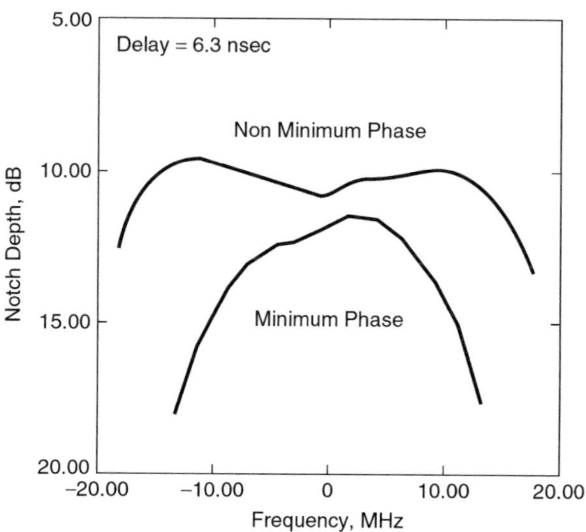

Fig. 3.26 Signature curves for a 16-QAM digital radio at 10^{-3} bit error rate. (From [7], with the permission of the IEEE)

frequencies (below and above the channel center frequency) where a notch of depth of a given value, normally 40 dB, causes no degradation to system BER performance. Signature depth is defined as the mean value of the measured notch depth within the signature bandwidth.

3.5.4   Cross-Polarization Discrimination Degradation due to Fading

In order to double the capacity on a given transmission frequency, signals can be transmitted simultaneously on both horizontal and vertical polarization. As indicated in Sect. 3.2.3, this can be accomplished by using parabolic antennas with dual feeds or via the use of an orthomode transducer. Ideally, no energy transmitted on one polarization would be intercepted by the receiving antenna feed adjusted for the reception of the signal on the other polarization. In practice, this is never the case, but under unfaded conditions, antenna systems can be deployed that result in the signal of the unwanted polarization being 30–40 dB below that of the intended polarization. This difference between wanted and unwanted signal levels is referred to as *cross-polarization discrimination (XPD)*. At these levels of XPD, the effect of interference on BER performance, even on relatively high-level modulation systems, is negligible. Unfortunately, deep fades resulting from adverse path conditions can reduce XPD to unacceptable levels. These conditions are many and include heavy rain and strong frequency-selective effects from refraction or ground reflection. The mechanisms causing XPD reduction are somewhat complex and will not be discussed here. It is of interest to note, however, that under conditions of multipath fading, XPD remains approximately constant for fade levels down to about 20 dB, but beyond that point, decreases dB for dB with the fade level [8]. The ITU provides a method to predict XPD outage due to clear air effects in Rec. ITU-R P.530-17 [5].

To minimize the impact of degraded XPD, modern digital wireless systems with cross-polarized signals employ adaptive *cross-polarization interference cancellers* (XPICs), the fundamentals of which are discussed in Chap. 6.

3.6 External Interference

Often radio systems are subjected to interference from unwanted signals. In a given frequency band, channels are normally assigned one beside the other with alternate channels being on alternate polarization, either horizontal or vertical. In some instances, where permitted by the licensing authority, operators double the capacity available from a given channel by operating on both horizontal and vertical polarization. Thus, a receiver operating at the end of a path may be subjected to interference from other channels over the same path. Also, depending on the geography of a path, it may be interfered with by channels on other paths of the same multi-hop system to which it belongs or channels on independent radio systems operating in the same general geographic area. Finally, it may be interfered with by satellite, radar, or any other types of systems radiating a signal at or very close to the receiver frequency. Because receivers usually have selectivity that increases rapidly outside the passband, only co-channel and adjacent channel signals normally pose a threat to system performance. Discrimination to an adjacent channel signal is achieved via receiver selectivity and the cross-polarization discrimination of the receiving antenna. Discrimination to a co-channel signal is only possible if it's cross-polarized, in which event the antenna provides some of the discrimination. Figure 3.27 shows examples of intra-system interference in a three-link system. A receiver at a repeater site, tuned to frequency f_1, is shown being interfered with by an adjacent channel signal transmitted on the same path as the desired signal, and two co-channel signals transmitted from adjacent paths. A receiver at an end terminal, tuned to frequency f_2, is shown being interfered with by an "overshoot" signal transmitted from two links away.

Radio manufacturers characterize their product's resistance to interference by the product's *threshold to interference (T/I) ratio*. It is usually provided for co-channel and adjacent channel and is normally the ratio in decibels of the interference level to the 10^{-6} BER static threshold level that degrades the BER static threshold 1 dB. It is measured with an interfering signal that is of the same type as the desired signal. Typically, the higher the modulation complexity, the more sensitive a receiver is to interference. In countries that keep a current database of all operating radio frequencies and signal radiated power, frequency search companies, utilizing T/I ratios, try to find for their client's frequencies that can be used in a specific path that will not interfere with or be interfered by other radio systems in the region. Thus, interference is a problem that's normally reviewed and hopefully eliminated or minimized at the planning stage.

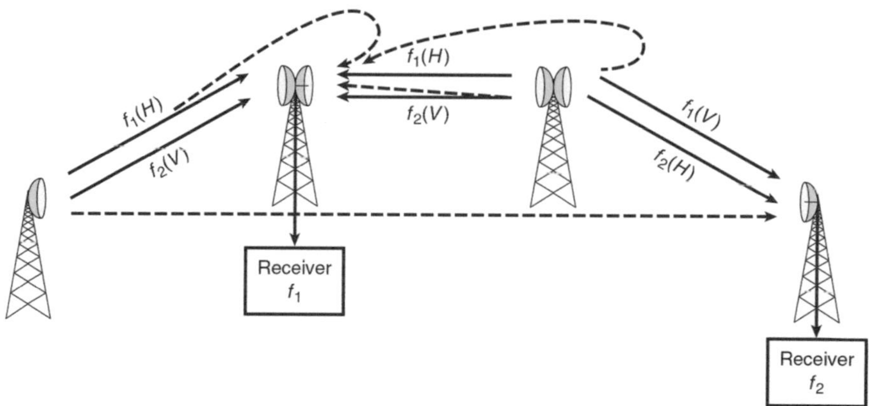

Fig. 3.27 Example of the adjacent channel and co-channel interference

3.7   Outage and Unavailability

An *outage event* is normally said to occur in a wireless receiver in any period of time when its BER exceeds 10^{-3}. Outage events where the BER exceeds 10^{-3} for periods of less than 10 s are referred to as periods of *unacceptable performance*. A loss of frame or synchronization also triggers an outage event. When outage events last for more than 10 consecutive seconds, traffic is usually disconnected, and a radio link is said to *unavailable*. The period of unavailability ends at the start of the period where the BER improves to better than 10^{-3} and stays better than 10^{-3} for ten consecutive seconds. Any one-second period where the BER exceeds 10^{-3} is termed a *severely errored second (SES)*. Any one second period where the bit error rate exceeds 10^{-6} but is less than 10^{-3} is termed a *burst errored second* (Burst ES). Any one second period where the *residual bit error rate (RBER)* exceeds 10^{-10} but is less than 10^{-6} is termed a *dribbling errored second* (Dribbling ES). Finally, when the RBER is less than 10^{-10}, the link is said to be in an *error-free state*.

Unavailability is meant to be a measure of the time that a system is not available for service. Long duration interruptions, such as those caused by equipment failure, as well as relatively short ones, such as those caused by rain fading or duct entrapment, contribute to unavailability. Unavailability objectives are normally expressed as a percentage of a year and refer to two-way system communications. Often, instead of specifying unavailability, *availability* is specified, availability being the time that a system is available for service. Thus, the availability percentage of a link is 100 minus the unavailability percentage.

Multipath-induced outages generally consist of periods lasting less than 10 s and are therefore contributors primarily to unacceptable performance. Objectives for unacceptable performance are defined unidirectionally and are usually measured over a one-month period in ITU regions and annually in North America. As multipath fading is seasonal, it results in a *worst month performance*. Rain-induced outages, on the other hand, generally consist of periods lasting minutes and are therefore contributors to unavailability.

Fixed wireless users in North America sometimes specify their link outage requirement in terms of *path reliability*. Path reliability is usually defined as the percentage of time annually that there is no outage due to fading. Thus, if P is the probability of outage due to fading, then the reliability, R, assuming outages are due only to fading, is given by:

$$R = (1 - P) \cdot 100\% \qquad (3.21)$$

Wireless transport links are often designed to meet a reliability of 99.999%. This translates to an outage time of only 5 min per year!

The ITU, in its recommendation P.530-17 [5], provides a procedure of predicting outage and hence the availability of an unprotected digital radio system due to multipath conditions. This procedure requires a number of computational steps and takes several parameters into account, including the systems signature width and signature depth (Sect. 3.5.3.3). The recommendation also provides a procedure to predict the decrease in outage via the use of diversity techniques such as space diversity (Sect. 3.8.1) and angle diversity (Sect. 3.8.2). It should be noted that [5] indicates that "Multipath fading and enhancement only need be calculated for path lengths longer than 5 km and can be set to zero for shorter paths." The recommended procedure in [5] for "quick planning applications" is given in Appendix B. Also given in Appendix B is an outage example for a 6 GHz, 28 MHz channel bandwidth, 4096-QAM, 20 km link which is shown to have a predicted availability of 99.999%.

The ITU also provides in [5] a procedure for predicting the outages and hence availability in the presence of hydrometeors such as rain, snow, hail, and fog. This procedure, like that covering multipath, is computationally quite demanding, but computer programs exist to simplify the process. A somewhat less accurate but computationally less demanding process, based in part on earlier ITU recommendations, is presented in Appendix C. This latter approach permits outage computations via a few relatively simple steps.

3.8   Diversity Techniques for Improved Reliability

Diversity is a means whereby two or more versions of the same simultaneously transmitted information are intercepted at a receiver site. It is used to improve a link's resistance to the degrading effect of multipath fading when the performance from the non-protected link is unacceptable. Three distinct types of diversity are employed in point-to-point, line-of-sight digital radio systems: *space diversity*, *angle diversity*, and *frequency diversity*. In space diversity, one signal is transmitted, but at the receiver, two or, in rare instances over difficult paths, three signals are received by using multiple antennas placed at differing vertical heights. Angle diversity is similar to space diversity but employs either two side-by-side receiving antennas at the same vertical height but adjusted at slightly different receiving directions or only one receiving antenna reflector fitted with two feed horns adjusted at slightly different angles. Either approach results in signals at the receivers that travel over slightly different paths. Finally, in frequency diversity, the same information is transmitted from one antenna but on two separate frequencies, and at the receiving end, one antenna intercepts these signals and feeds two separate receivers that demodulate the signals. For the more efficient use of spectrum, frequency diversity is normally avoided in favor of space diversity, angle diversity, or a combination of both. Diversity works on the principle that signals received via different physical or electrical paths will be largely uncorrelated with respect to multipath-induced distortion. Thus, by clever choices at the receiving end in how the original data stream is reconstructed, most errors resulting from multipath fading can be eliminated. Improvement in outage time on a radio link resulting from diversity is expressed as a *diversity improvement factor, I*, which is defined as the ratio of outage time without diversity to the outage time with diversity. Methods for predicting outage probability for space, angle, and frequency diversity digital radio systems, and such systems employing a combination of space and frequency diversity are given in Rec. ITU-R P.530-17 [5].

3.8.1   Space Diversity

Space diversity is helpful in combating both flat fading as well as frequency selective fading. Figure 3.28a shows a simplified one-way vertical space diversity link. In such links, a number of different techniques are used at the receiving end to minimize the degradation in error rate performance relative to the unprotected link. These techniques vary from combining the received signal at RF to combining the received signals at IF to selecting the appropriate data at baseband. However, RF combining is rarely used due to its high cost relative to IF combining. Figure 3.28b shows an example of space diversity RF combining, Fig. 3.28c shows IF combining, and Fig. 3.28d shows switching at baseband. Switching, either at RF or IF, is not employed on digital radio systems because of the large number of data interruptions that such an approach would result in during fading. However, as will be discussed below, switching at baseband can be accomplished without data interruption. Combining either at RF or IF minimizes the amount of receiver hardware required as common equipment is used after the combiner. Baseband switching, however, requires two complete receivers to drive the switch. Following is a summary of some of the space diversity combining methods used on broadband digital links:

The *maximum power combiner (MPC)* [14], also called the *equal gain combiner (EGC)*, was initially developed for analog systems, was among the first used in digital space diversity systems and continues to be widely used. It is a device that can operate at either RF or IF. It shifts the phase of the signal entering through one antenna before combining it, with equal weighting, with the signal entering through the other. The amount of phase shift applied is such as to bring together in phase the

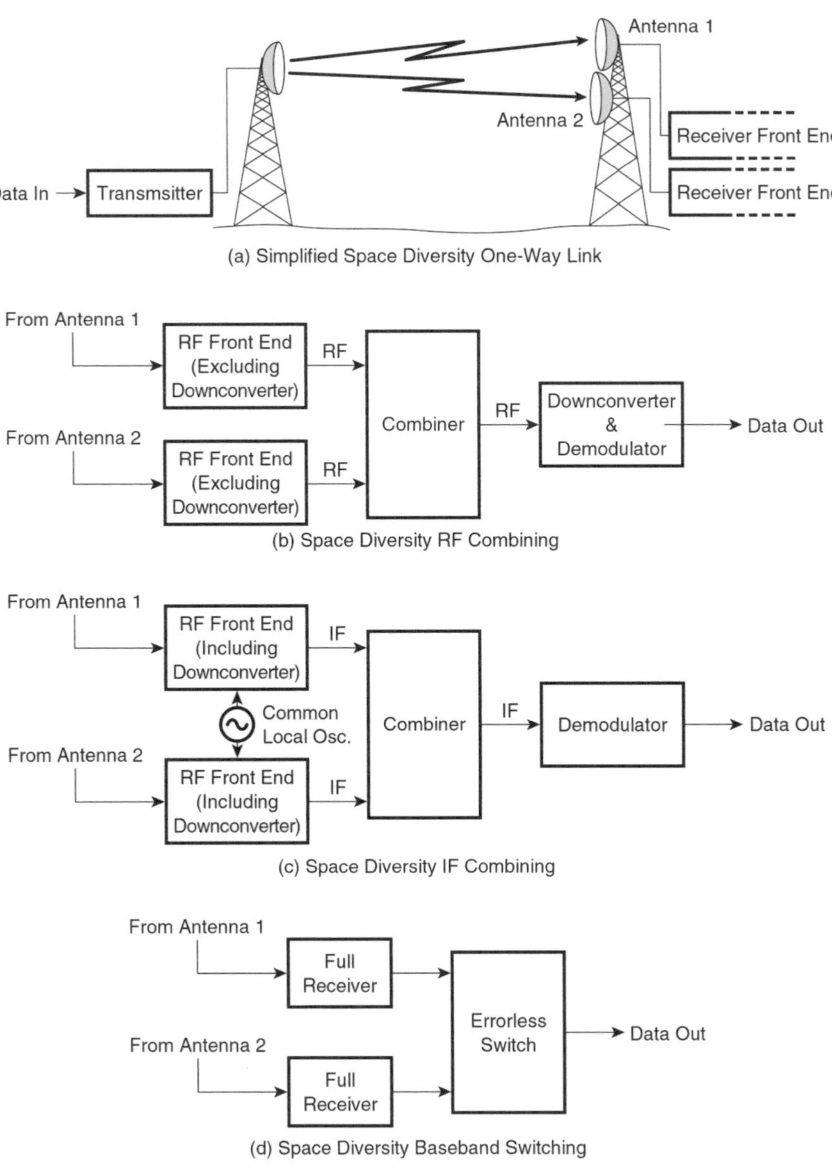

Fig. 3.28 Space diversity configurations

center frequency spectral components of the two received signals. The process is called co-phasing. As a result, the power of the combined received signal at the channel center frequency is maximized. Certain field results of outage times resulting from applying this method showed a factor of reduction that varied between one-half and one and one-half orders of magnitude relative to the non-diversity case [15].

The *maximal ratio combiner* (MRC), like the MPC, co-phases and sums, at RF or IF, the incoming signals. However, before summing, it weights the signals in an adaptive fashion so as to maximize its instantaneous output signal-to-noise ratio. As one would suspect, it provides somewhat better performance relative to the MPC.

The *minimum dispersion combiner* (MDC) [16] also an RF or IF type, combines the received signals in such a fashion as to minimize the in-band amplitude dispersion. It achieves this by monitoring the power present in two or more narrow bands within the combined channel bandwidth and adjusting the relative phase of the two received signals based on the monitored results. Unfortunately, if a fading situation should produce input signals at the two antennas that are similar over the frequency range of the channel, the signals will mutually cancel, leaving no useful combined output. This is avoided by switching to the maximum power combining state under such circumstances. Though more complex than the pure maximum power combiner, the minimum dispersion combiner is particularly attractive in high capacity and hence high bandwidth digital radio systems due to its superior minimization of in-band dispersion. Certain field results indicated diversity improvement factors of more than an order of magnitude when this method of combining was employed [16].

The *diversity baseband switch* switches between the two-receiver baseband outputs and on space diversity systems is usually designed to perform error-free switching. In such switching, the performance of each receiver is constantly monitored and when distortion to amplitude, slope, etc. is detected on one stream the "anticipatory" switch selects data from the other stream before errors occur on either channel. It achieves errorless switching by using elastic storage to line up each data stream so that the same bit is available from either stream at exactly the same time. This way, no extra bits are added, and neither are bits deleted.

3.8.2   Angle Diversity

Angle diversity like space diversity is helpful in combating both flat fading and frequency selective fading and can be employed in cases where adequate space diversity is not possible or to reduce tower heights. It can also be used in conjunction with space diversity by tilting the antennas at different upward angles. Early researchers into space diversity found that the diversity improvement factor did not increase linearly with separation, but rather seemed to be independent of practical separation distances. At about the same time researchers began experiments with different type diversity antennas at the same height and discovered large diversity improvement factors [17]. Further research led to the finding that by slightly offsetting the elevation adjustments of two identical antennas at the same height, or by slightly offsetting the elevation adjustments of two feed horns of a single parabolic reflector, similar large diversity improvement factors were realized. This approach was therefore appropriately named angle diversity. The reason that such small differences in the composite signal received by the feed horns can produce a large diversity improvement factor is as follows. The creation of a notch deep enough to cause a digital radio outage requires that the two predominant combined rays be of very nearly equal power and 180° out of phase at the notch frequency. However, in this condition, even a small reduction in the power level of one ray relative to the other can reduce the notch depth significantly. For example, a 1 dB reduction in the relative power level can reduce notch depth from 40 dB to less than 20 dB [17]. Thus, during multipath fading, if the received signal on one feed has a deep notch, it is unlikely the signal on the other feed will simultaneously have a deep notch, since it is receiving signal components of slightly different magnitude. Angle diversity using one reflector significantly lowers the cost of diversity implementation relative to vertical separation diversity as it not only lowers the antenna cost but also reduces antenna wind load on the supporting tower and potentially lowers the tower height. Angle diversity is normally employed where installation limitations such as space, aesthetics, and tower loading prohibit space diversity. Like space diversity systems, angle diversity systems employ maximum power combiners, minimum dispersion combiners, dispersion-sensing IF combiners, and diversity baseband switches.

3.9 Summary

This chapter addressed the fixed wireless line-of-sight path, the transmission media for 5G wireless transport. It examined the behavior of such paths, looking at many influencing factors. Such factors include antenna characteristics, free space propagation, and fading phenomena as influenced by atmospheric effects such as multipath and rain attenuation and terrain effects such as diffraction and reflection. Finally, external interference, outage, availability, and diversity techniques for improved availability were reviewed.

References

1. Kraus JD (1950) Antennas. McGraw-Hill Book Company, Inc, New York
2. Rogers DV, Olsen RL (1976, November) Calculation of radiowave attenuation due to rain at frequencies of up to 1000 GHz, CRC Report No. 1299. Communications Research Centre, Department of Communications, Ottawa
3. ITU Recommendation ITU-R P.838-3 (2005) Specific attenuation model for rain use in prediction methods. ITU, Geneva
4. ITU Recommendation ITU-R P.676-12 (2019) Attenuation by atmospheric gases. ITU, Geneva
5. ITU Recommendation ITU-R P.530-17 (2017) Propagation data and prediction methods required for the design of terrestrial line-of-sight systems. ITU, Geneva
6. ITU Recommendation ITU-R P.526-7 (2001) Propagation by diffraction. ITU, Geneva
7. Rummler WD, Coutts RP, Liniger M (1986, November) Multipath fading channel models for microwave digital radio. IEEE Commun Mag 24(11):30–42

8. Ivanek F (ed) (1989) Terrestrial digital microwave communications. Aertech House, Inc., Norwood
9. Members of the technical staff, Bell Telephone Laboratories (1971) Transmission systems for communications, Revised Fourth Edition, Bell Telephone Laboratories, Inc
10. Morais DH (2004) Fixed broadband wireless communications. Prentice Hall PTR, Upper Saddle River
11. Siller CA Jr (1984, February) Multipath Propagation: Its associated countermeasures in digital microwave radio. IEEE Comm Mag 22(2)
12. Anderson CW, Barber SG, Patel RN (1979, December) The effect of selective fading on digital radio. IEEE Trans Commun COM-27:1870–1876
13. Rummler WD (1979, May/June) A new Selective Fading Model: application to propagation data. Bell Syst Tech J 58(5):1037–1071
14. Rummler WD (1984) Modeling the diversity performance of digital radios with maximum power combiners, IEEE conference on communications, pp 657–660
15. Documents CCIR Study Group Period 1982–1986, Effects of propagation on the design and operation of line-of-sight radio-relay systems, Draft Report, 784-1, (Annex 1. Docs, 9/2049/402)
16. Komaki S, Tajima K, Okamoto Y (1984, April) A minimum dispersion combiner for high capacity digital microwave radio. IEEE Trans Commun COM-32(4):419–428
17. Lin SH, Lee TC, Gardina MF (1988, February) Diversity protections for digital radio – summary of ten-year experiments and studies. IEEE Commun Mag 26(2):51–64

Chapter 4
Digital Modulation: The Basic Principles

4.1 Introduction

The modulation methods used in most fifth generation (5G) wireless transport links are largely various versions of quadrature amplitude modulation (QAM). At the low end of modulation order, we find BPSK, which is not QAM, followed by QPSK, which is really 4-QAM. At the high end, we find 8192-QAM and in a few instances 16,384-QAM, referred to also as 16 K-QAM. The QAM modulation order can be 2^{2n} or 2^{2n+1} where n is an integer of value 1 or above. These are the methods, therefore, that we are going to study in this chapter, along with a number of schemes involved in their realization such as up- and downconversion and carrier and symbol timing recovery. In these methods, baseband signals linearly modulate an RF carrier. Ideally, we would like to cut to the chase and study these methods immediately. This is not what we will do, however. First, the fundamentals of baseband transmission techniques will be reviewed, as these form the foundation upon which the techniques involving the linear modulation of an RF carrier are based. This approach will allow us to gain a strong intuitive understanding of digital modulation and the spectral efficiency and bit error rate characteristics thereof.

4.2 Baseband Data Transmission

Data transported by communication systems is typically random or very close to random in nature. A stream of random, bipolar (positive and negative), full-length rectangular pulses (a full-length rectangular pulse is one of constant height for the entire pulse duration), where each pulse is equiprobable (probability of being positive is the same as being negative), is in fact a string of nonperiodic pulses. This is so because the valuc (polarity) of each pulse is entirely independent of all other pulses in the stream. For such a stream, of pulse duration τ, and hence symbol

D. H. Morais, *5G and Beyond Wireless Transport Technologies*,
https://doi.org/10.1007/978-3-030-74080-1_4

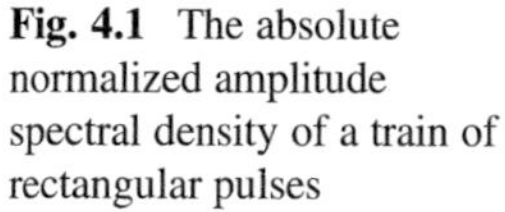

Fig. 4.1 The absolute normalized amplitude spectral density of a train of rectangular pulses

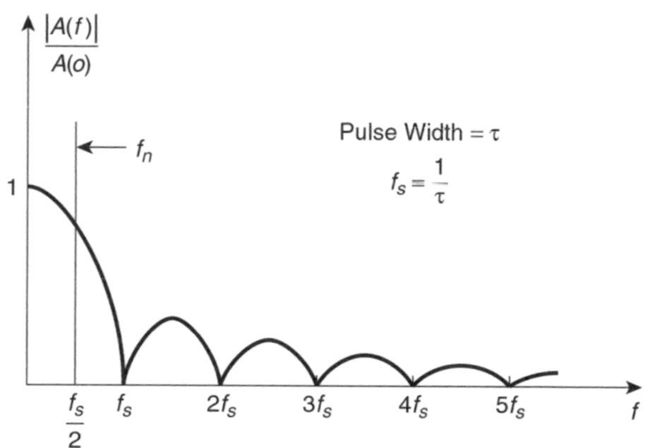

rate $f_s = 1/\tau$ symbols per second, the two-sided amplitude spectral density is that of each of its nonperiodic pulses and thus as given in Eq. (D. 3). The absolute value of its normalized positive (real) side is shown in Fig. 4.1. It occupies infinite bandwidth, with the first null at f_s. The spectrum within the first null is normally referred to as the main lobe. In communication systems, bandwidth is normally at a premium. Thus, the designer is motivated to filter the transmitted signal down to the minimum bandwidth possible without introducing errors into the transmission. Further, at the receiver, the incoming signal is normally filtered to minimize the negative effects of noise and interference. Filtering the signal results in changes to the shape of the original pulses, spreading their energy into adjacent pulses. This spreading effect is known as *dispersion* and can result in distortion of the pulse amplitude at the sampling instant unless carefully controlled (the sampling instant is the instant at which the receiver decides on the polarity and integer value of the received pulse). In the *Nyquist criterion* on bandwidth transmission, it is shown in [1] that the minimum real channel bandwidth that independent symbols at a rate f_s can be transmitted through, without resulting in symbol amplitude distortion at the sampling instant, is the Nyquist bandwidth $f_n = f_s/2$. Thus, for the rectangular pulse stream described above, the minimum transmission bandwidth is half the width of the main lobe. If the stream is binary, one transmitted symbol contains one information bit. Thus, the bit rate f_b of the stream is the same as the symbol rate f_s and the minimum real transmission bandwidth f_n is $f_b/2$.

In a *pulse amplitude modulated* (PAM) digital baseband system, bits are transmitted as symbols. If the transmitted stream is a two-level one, then the symbols are the bits. If, however, the transmitted stream is a 2^n level one, where $n \geq 2$, incoming bits are converted to symbols. Let us consider the case of a four-level PAM stream. Here, two information bits are encoded into each symbol, the four symbol levels being encoded to represent 00, 01, 10, and 11. As a result, the symbol rate f_s is $f_b/2$. In general, for L-level transmission systems, each transmitted symbol contains n information bits, where n is given by $n = \log_2 L$. Thus, $f_s = f_b/n$, and the minimum transmission bandwidth $f_n = f_b/2n$. For example, for an eight-level PAM signal, $n = 3$ and thus $f_n = f_b/6$..

We will initially look at the transmission of impulses, then at the more practical case of pulse transmission. For impulse transmission, an ideal low pass brickwall filter, of bandwidth $f_s/2$ and fixed time delay, will result in a nondistorted pulse amplitude at the sampling instant. Unfortunately, such a filter is unrealizable. Nyquist has shown, however, that if the amplitude characteristic of the brickwall filter is modified to have a gradual roll-off, with odd symmetry about the Nyquist bandwidth f_n, then a non-distorted pulse amplitude at the sampling instant is preserved. One class of such a filter that is most often used in digital communication systems is the so-called *raised cosine filter*. The amplitude characteristic of such a filter for impulse transmission, $H_{im}(f)$, consists of a flat portion followed by a roll-off portion that has a sinusoidal form. It is characterized in terms of its *roll-off factor* α, where α is defined as f_x/f_n, f_x being the amount of bandwidth used in excess of the Nyquist bandwidth. The roll-off factor α may vary between 0 and 1, corresponding to an excess bandwidth of 0–100%. The amplitude characteristic $H_{im}(f)$ is given mathematically by

$$
\begin{aligned}
H_{im}(f) \quad &= 1, & 0 &< f < f_n - f_x \\
&= \frac{1}{2}\left[1 - \sin\frac{\pi}{2\alpha}\left(\frac{f}{f_n} - 1\right)\right], & f_n - f_x &< f < f_n + f_x, \qquad (4.1)\\
&= 0, & f &> f_n + f_x
\end{aligned}
$$

where $\alpha = \frac{f_x}{f_n}$

Figure 4.2, which is from Feher [1], is a graphical representation of $H_{im}(f)$.

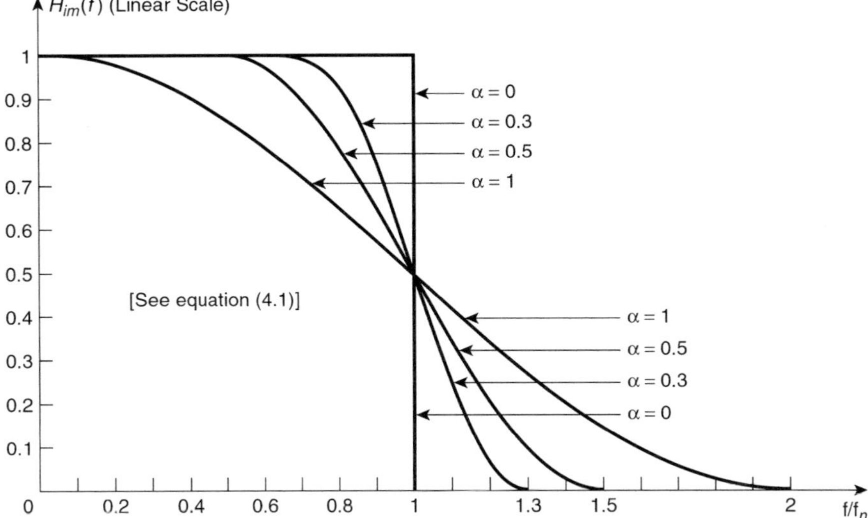

Fig. 4.2 Amplitude characteristics of the Nyquist channel for impulse transmission. (From [1], with the permission of the author)

The phase characteristic $\varphi(f)$ of the raised cosine filter is linear over the frequency range where the amplitude response is greater than zero and is thus given by

$$\phi(f) = Kf \quad 0 < f < f_n + f_x \tag{4.2}$$

where K is a constant.

Because the input to the filter defined above is a stream of impulses, then the amplitude spectral density of each impulse at the filter input is of constant amplitude as per Eq. (D.5), and as per Eq. (D.13), the amplitude spectral density at the filter output $S_{rc}(f)$ has the identical characteristic of the filter transfer function $H_{im}(f)$, i.e.,

$$S_{rc}(f) = H_{im}(f) \tag{4.3}$$

A subtle but important point to note is that it is this output spectral density and hence the received pulse shape that results in nondistorted pulse amplitudes at the sampling instants, not the filter transfer function per se.

In practical systems, pulses of finite duration, not impulses, are used for digital transmission. A commonly used pulse is the full-length rectangular pulse. We note from Eq. (D.3) that the spectrum of such a pulse has a (sin x)/x form. For nondistorted pulse amplitudes at the sampling instant, we desire that the spectral density and hence the pulse shape, at the filter output of a transmission system conveying such pulses, be the same as that for the impulse case discussed above, namely, $S_{rc}(f)$. To achieve this thus requires that the transfer function of the low-pass filter, $H_{rp}(f)$ say, be the filter transfer function for impulses modified by the factor x / (sin x). Thus $H_{rp}(f)$ is given by

$$H_{rp}(f) = \frac{\pi f/2f_n}{\sin\left(\pi f/2f_n\right)} H_{im}(f) \tag{4.4}$$

Figure 4.3, which is also from Feher [1], is a graphical representation of $H_{rp}(f)$.

The general half-sided shape of time responses resulting from filtering a pulse to achieve a raised cosine spectral density $S_{rc}(f)$ as defined in Eq. (4.3) is shown in Fig. 4.4 for excess bandwidth values α of 0% and 50%. The full-sided time response extends from the negative to positive time axis with the pulse centered at time 0. Following pulses would be centered at time 1, 2, etc. The shape of these time responses, irrespective of roll-off factor, explains why there is no pulse amplitude distortion at the sampling instant as a consequence of the spreading of adjacent pulses. Pulses are sampled at their maximum amplitude. However, with a raised cosine spectrum, spread pulses adjacent to a pulse being sampled are always at zero amplitude at the time of sampling.

The time responses shown in Fig. 4.4 are for perfect filters. In practice, raised cosine filters are never perfect. As a result, spread pulses adjacent to the pulse being sampled are not necessarily zero amplitude at the time of the sampling. Consequently, the sampled pulse is not always at its nominal value but may be of a value

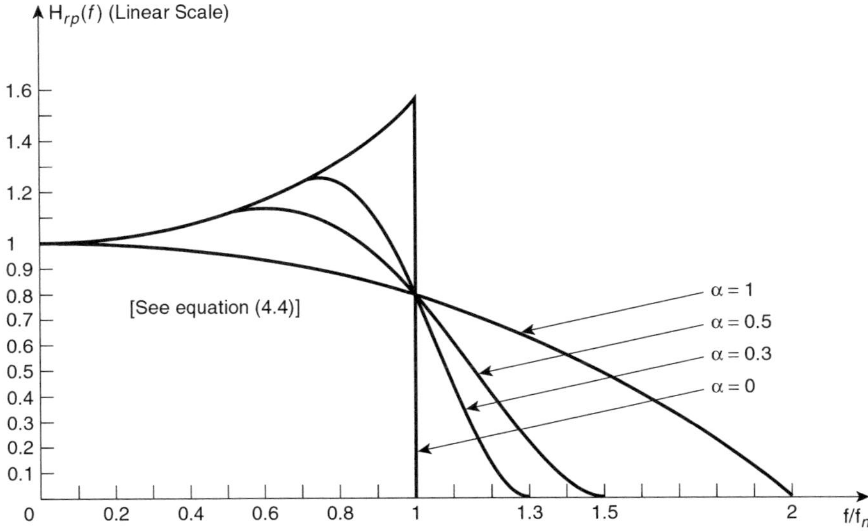

Fig. 4.3 Amplitude characteristics of the Nyquist channel for rectangular pulse transmission. (From [1], with the permission of the author)

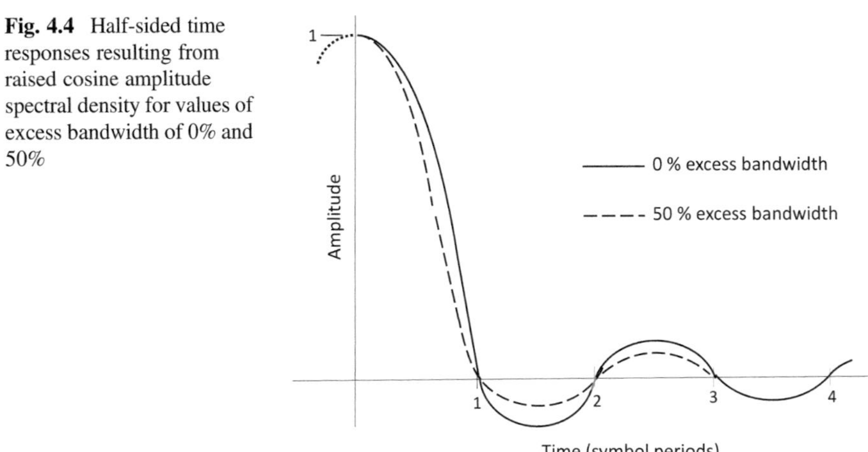

Fig. 4.4 Half-sided time responses resulting from raised cosine amplitude spectral density for values of excess bandwidth of 0% and 50%

that is more or less than expected. When this occurs, it is referred to as *intersymbol interference (ISI)*.

In the laboratory, intersymbol interference can be observed with the aid of an oscilloscope. On the vertical scale, one displays the filtered response to the random pulse sequence under study, and the horizontal time base is set to the symbol duration. The resulting display is referred to as an *eye diagram* or an *eye pattern*. The inherent persistence of the oscilloscope's cathode-ray tube results in the display of superimposed pulse responses. Figure 4.5a shows what an eye pattern would look

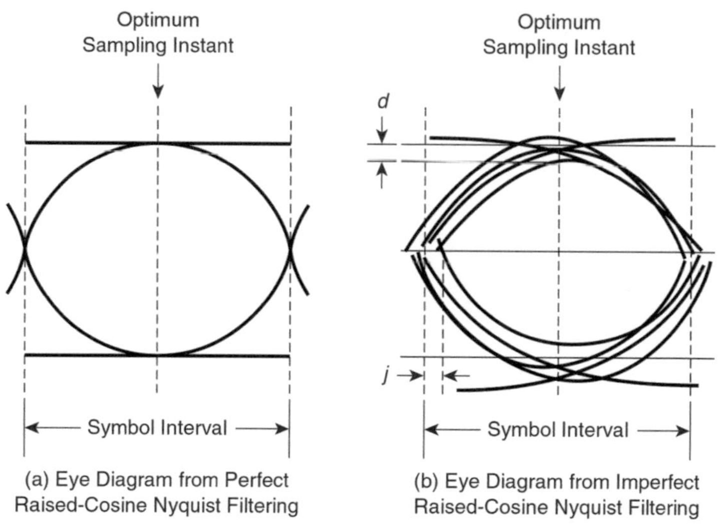

Fig. 4.5 Eye diagrams from raised-cosine Nyquist filtering of a binary stream of rectangular pulses

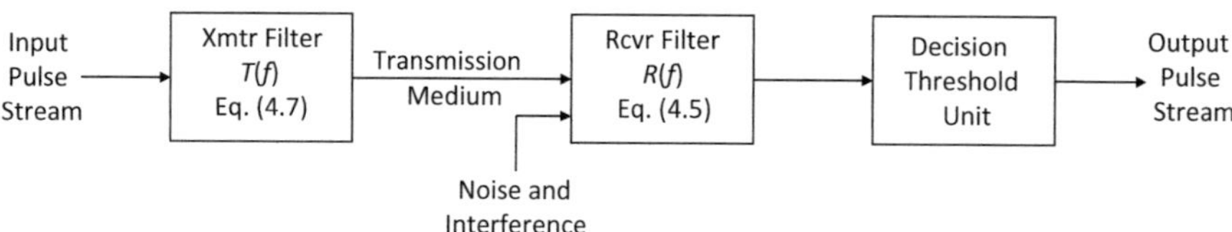

Fig. 4.6 Basic baseband digital transmission system

like that results from unlimited traces created by perfect raised cosine filtering of a binary stream of full-length rectangular pulses. Figure 4.5b shows what an eye pattern would look that results from a few symbol traces created by imperfect filtering of a binary stream of full-length rectangular pulses. In Fig. 4.5b, the difference d shown between the nominal peak amplitude and the minimum peak amplitudes is a measure of distortion caused by ISI, and the difference j between the nominal zero crossing time and the furthest removed zero crossing times is a measure of the *timing jitter*. The larger the ISI, the worse the error rate performance of the system. The larger the timing jitter, the larger the symbol sampling clock jitter, since the sampling clock is recovered via special circuitry from the incoming signal (Sect. 4.5.3). Jitter on the sampling clock can lead to degraded error rate performance.

The essential components of a baseband digital transmission system are shown in Fig. 4.6. The input signal can be a binary or multilevel (>2) pulse stream. The transmitter low-pass filter, with transfer function $T(f)$, is used to limit the transmitted spectrum. Noise and other interference are picked up by the transmission medium and fed into the receiver filter. The receiver filter, with transfer function $R(f)$, minimizes the noise and interference relative to the desired signal. The output of the receiver filter is fed to a decision threshold unit which, for each pulse received, decides what was its most likely original level and outputs a pulse of this level. For a binary pulse stream, it outputs a pulse of amplitude $+V$ say, if the input pulse is equal to or above its decision threshold, which is 0 volts. If the input pulse is below 0 volts, it outputs a pulse of amplitude $-V$.

In designing the system, a non-distorted pulse amplitude at the sampling instant is desirable at the output of the receive filter. This is normally achieved by employing filtering that results in a raised cosine amplitude spectral density S_{rc} at the receiver filter output. Thus, for full-length rectangular pulses, the combined transfer function of the transmitter filter and receiver filter (assuming that the transmission medium results in negligible impairment) should be as given in Eq. (4.4). There are an infinite number of ways of partitioning the total filtering transfer function between the transmitter filter and the receiver filter. Normally, however, the receiver filter is chosen to maximize the signal-to-noise ratio at its output as this optimizes the error rate performance in the presence of noise. It can be shown [2, 3] that, for white Gaussian noise, the receiver filter transfer function $R(f)$ that accomplishes this is given by

$$R(f) = |S_{rc}(f)|^{\frac{1}{2}} \tag{4.5}$$

A filter with such a transfer function is referred to as a *root-raised cosine* (RRC) filter.

The transmitter filter transfer function $T(f)$ is then chosen to maintain the desired composite characteristic, i.e.,

$$T(f).R(f) = H_{rp}(f) \tag{4.6}$$

Thus, by Eqs. (4.4), (4.5), and (4.6),

$$T(f).|S_{rc}(f)|^{\frac{1}{2}} = \frac{\pi f/2f_n}{\sin\left(\pi f/2f_n\right)} S_{rc}(f)$$

and hence

$$T(f) = \frac{\pi f/2f_n}{\sin\left(\pi f/2f_n\right)} |S_{rc}(f)|^{\frac{1}{2}} \tag{4.7}$$

Figure 4.7 shows plots of $R(f)$ and $T(f)$ for $\alpha = 0.5$.

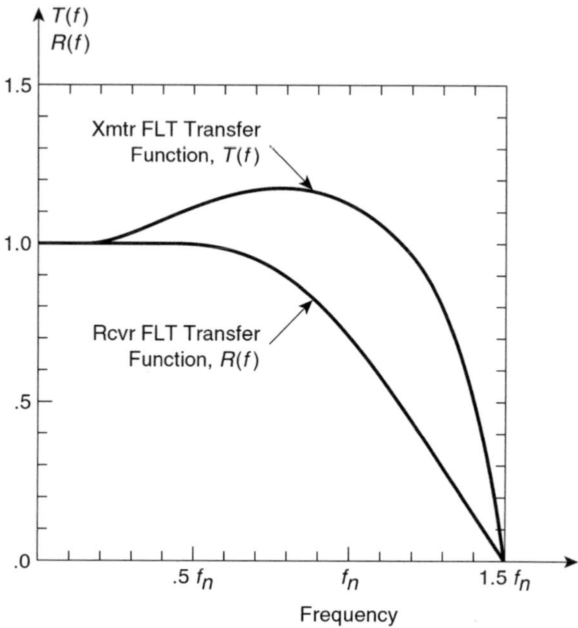

Fig. 4.7 Amplitude transfer functions of transmitter and receiver filters when $\alpha = 0.5$

4.3 Linear Modulation Systems

Above we discussed PAM baseband systems. Wireless communication systems, however, operate in assigned frequencies that are considerably higher than baseband frequencies. It is thus necessary to employ modulation techniques that shift the baseband data up to the operating frequency. In this section, we consider linear modulation systems. These systems are so-called because they exhibit a linear relationship between the baseband signal and the modulated RF carrier. As a result of this relationship, their performance in the presence of noise and other impairments can be deduced from their equivalent baseband forms, hence our earlier review of baseband systems. We will commence this study by reviewing so called *double-sideband suppressed carrier (DSBSC)* modulation as this modulation forms the foundation on which many of the most widely used linear modulation methods are based.

4.3.1 Double-Sideband Suppressed Carrier (DSBSC) Modulation

A simplified DSBSC system for PAM signal transmission is shown in Fig. 4.8. First, a polar L-level PAM input signal, $a(t)$, with equiprobable symbols, is filtered with the low-pass filter, F_T, to limit its bandwidth to f_m say, and the filtered signal $b(t)$

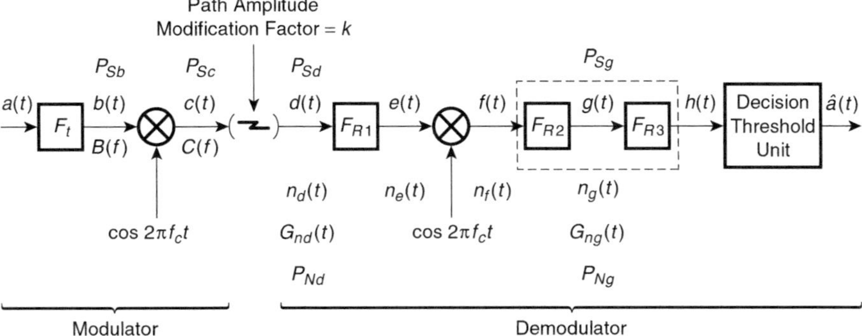

Fig. 4.8 Simplified one-way DSBSC system for PAM transmission

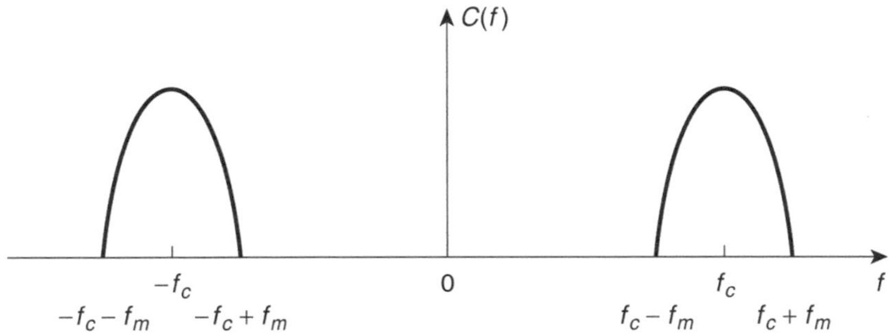

Fig. 4.9 DSBSC signal amplitude spectral density representation

applied to a multiplier. Also feeding the multiplier is a sinusoidal signal at the desired carrier frequency, f_c. As a result, the output signal of the multiplier, $c(t)$, is given by

$$c(t) = b(t) \cos 2\pi f_c t \qquad (4.8)$$

If the amplitude spectral densities of $b(t)$ and $c(t)$ are represented as $B(f)$ and $C(f)$ respectively, then, by Eq. (D.9), $C(f)$ is given by

$$C(f) = \frac{1}{2}B(f + f_c) + \frac{1}{2}B(f - f_c) \qquad (4.9)$$

Thus, as shown in Fig. 4.9, $C(f)$ consists of two spectra. One is real centered at f_c, the other imaginary centered at $-f_c$, and each has a bandwidth $2f_m$ and an amplitude half that of $B(f)$. As these spectra are each symmetrically disposed on either side of the carrier frequency, the signal is referred to as a *double-sideband (DSB)* signal. Further, as $b(t)$ consists of bipolar equiprobable symbols and thus an average value of zero, it has no nonzero fixed-level component. As a result, $c(t)$ contains no discrete carrier frequency component and is thus referred to as a *suppressed carrier* signal.

We assume that $c(t)$ travels over a linear transmission path and arrives at the demodulator input modified in amplitude by the factor k. Thus, the input signal $d(t)$ to the receiver is given by

$$d(t) = k \cdot c(t) \tag{4.10}$$

The received signal $d(t)$ is passed through the bandpass filter F_{R1} to limit noise and interference. The bandwidth W of F_{R1} is normally greater than $2f_m$, the bandwidth of $d(t)$, so as to not impact the spectral density of $d(t)$. Assuming this to be the case, the output signal $e(t)$ of F_{R1} is given by

$$e(t) = d(t) \tag{4.11}$$

The signal $e(t)$ is fed to a multiplier. Also feeding the multiplier is the sinusoidal signal $\cos 2\pi f_c t$. As a result, the output of the multiplier $f(t)$ is given by

$$f(t) = e(t) \cos 2\pi f_c t \tag{4.12}$$

Substituting Eqs. (4.11), (4.10), and (4.8) into Eq. (4.12) we get

$$\begin{aligned}
f(t) &= k \cdot b(t) \cos^2(2\pi f_c t) \\
&= \frac{k}{2} b(t)[1 + \cos(2 \cdot 2\pi f_c t)] \\
&= \frac{k}{2} b(t) + \frac{k}{2} b(t) \cos(2 \cdot 2\pi f_c t)
\end{aligned} \tag{4.13}$$

Thus, by multiplying $e(t)$ by $\cos 2\pi f_c t$, a process referred to as *coherent detection*, we recover $b(t)$ and create a second signal with the same double-sided bandwidth as $b(t)$ but centered at $2f_c$. The signal $f(t)$ is fed into the low-pass filter F_{R2} that eliminates the component of the signal centered about $2f_c$ while leaving the baseband component undisturbed. Thus, the output of F_{R2}, $g(t)$, is given by

$$g(t) = \frac{k}{2} b(t) \tag{4.14}$$

The signal $g(t)$ is fed to F_{R3} for final pulse shaping prior to level detection in the decision threshold unit. In practice, F_{R2} and F_{R3} are combined into one but are shown separately here to add clarity to the analysis. The output, $\hat{a}(t)$, of the decision threshold unit is a PAM signal that is the demodulator's best estimate of the modulator input signal, $a(t)$.

4.3.2 *Binary Phase-Shift Keying (BPSK)*

A special case of PAM transmission via a DSBSC system is when the PAM signal a (t) in Fig. 4.8 has a binary, polar format. In this situation, if the filtered signal $b(t)$ has maximum peak amplitude of $\pm b$ volts say, then the modulated signal $c(t)$ varies between $c_1(t)$ and $c_0(t)$ as $b(t)$ varies between $+b$ and $-b$, where

$$c_1(t) = b \cos 2\pi f_c t \tag{4.15}$$

$$c_0(t) = -b \cos 2\pi f_c(t)$$
$$= b \cos (2\pi f_c t + \pi) \tag{4.16}$$

When $b(t)$ is positive, the phase of $c(t)$ relative to the carrier phase is 0°. When $b(t)$ is negative, the phase of $c(t)$ relative to the carrier phase is π radians or -180°. Thus, the relative phase has only two states. This modulation is referred to as *binary phase-shift keying* (BPSK) and represents the simplest linear modulation scheme. Figure 4.10 shows typical examples of signals $a(t)$, $b(t)$, and $c(t)$. Figure 4.11 shows the *signal space* or *vector* or *constellation diagram* of $c(t)$. This diagram portrays both the amplitude and phase of $c(t)$ at the instances when the modulating signal $b(t)$ is at its peak.

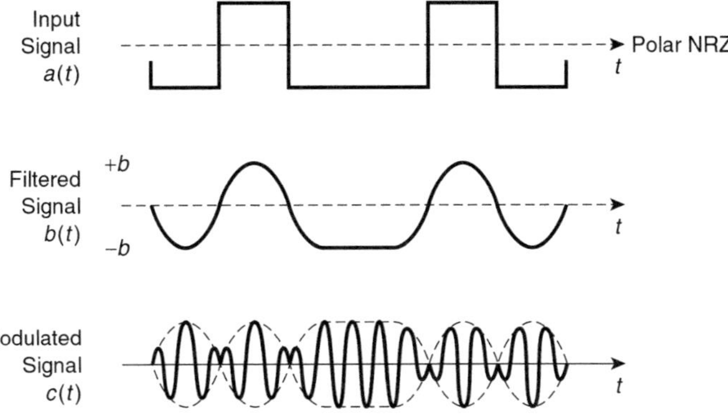

Fig. 4.10 Typical BPSK signals

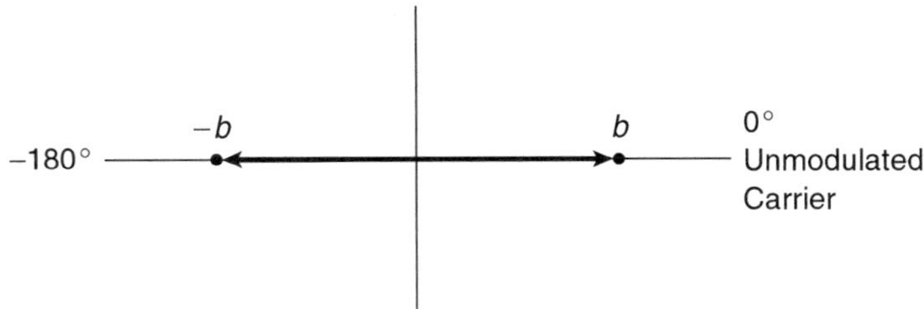

Fig. 4.11 Signal space diagram of BPSK modulated signal

It can be shown [5] that the probability of bit error $P_{be(BPSK)}$ of a BPSK system, with optimum filtering, and in the presence of white Gaussian noise, is given by

$$P_{be(BPSK)} = Q\left[\left(\frac{2P_S}{P_N}\right)^{\frac{1}{2}}\right] \tag{4.17}$$

where $P_S = $ the average signal power at the demodulator input,
and $P_N = $ the noise power in the two-sided Nyquist bandwidth at the demodulator
   input,
and $Q(x)$ is the Q function which is related to the well-known complementary error
   function $erfc(x)$ and for which tabulated values are available.

In Eq. (4.17) above, probability of error is defined in terms of a signal-to-noise ratio. In digital communication systems, however, it is just as common to define the probability of bit error,P_{be}, in terms of the ratio of the energy per bit E_b in the received signal to the noise power density N_0 (Watts/Hz) at the receiver input. Defining P_{be} in terms of E_b/N_0 makes it easy to compare the error performance of different modulation systems for the same bit rate. Given that, for a BPSK system with bit rate f_b and hence a single-sided, double sideband Nyquist bandwidth also of f_b,

$$P_S = E_b \cdot f_b \tag{4.18}$$

and

$$P_N = N_0 f_b \tag{4.19}$$

Thus

$$\frac{P_S}{P_N} = \frac{E_b}{N_0} \tag{4.20}$$

and hence

$$P_{be(BPSK)} = Q\left[\left(2\frac{E_b}{N_0}\right)^{\frac{1}{2}}\right] \tag{4.21}$$

Figure 4.12 shows the power spectral density of BPSK when the modulating signal is unfiltered. This power spectral density has the same $\sin x/x$ form as that of the two-sided baseband signal, except that it is shifted in frequency by f_c. It can be shown [4] to be given by

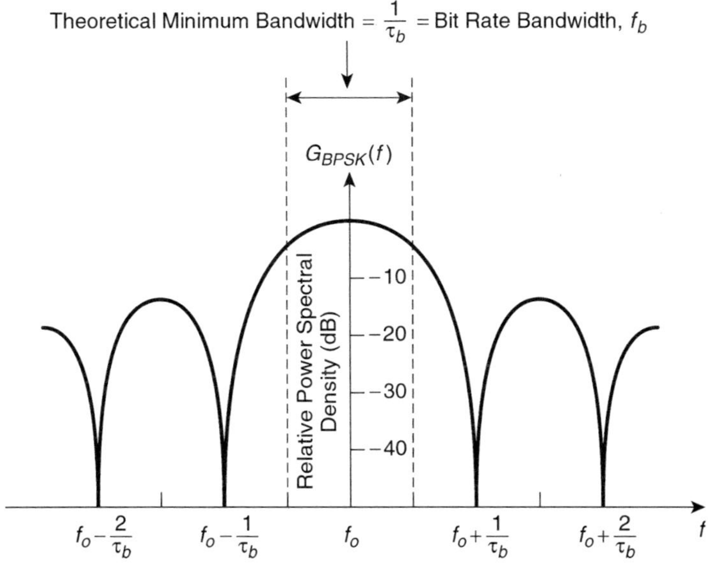

Fig. 4.12 Power spectral density of BPSK

$$G_{BPSK}(f) = P_S\tau_b \left[\frac{\sin \pi(f - f_c)\tau_b}{\pi(f - f_c)\tau_b}\right]^2 \tag{4.22}$$

where τ_b, which is equal to $1/f_b$, is the bit duration of the baseband signal.

Also shown in Fig. 4.12 is the single-sided, i.e., the real side only, double sideband Nyquist bandwidth of the system, which is equal to f_b. Thus, at its theoretical best, BPSK is capable of transmitting 1 bit per second in each Hertz of transmission bandwidth. The system is therefore said to have a maximum *spectral efficiency* of 1 bit/sec/Hz. Because, as indicated earlier, filtering to 0% excess bandwidth to achieve the Nyquist bandwidth is not practical, real BPSK systems have spectral efficiencies less than 1 bit/sec/Hz. For an excess bandwidth of 25%, then data at a rate of 1 bit/sec requires 1.25 Hertz of bandwidth, leading to a spectral efficiency of 0.8 bits/sec/Hz.

As we shall see in succeeding sections, spectral efficiencies much greater than that afforded by BPSK are easily realizable. As a result, BPSK is rarely used in wireless communication networks, where, as a rule, the available spectrum is limited and thus highly valued. Nonetheless, an understanding of its operating principles is very valuable in analyzing *quadrature phase-shift keying* (QPSK), a popular modulation technique.

For BPSK modulated with a pulse train of rectangular pulses, for example, signal $a(t)$ shown in Fig. 4.10, the amplitude of the modulated signal is unchanged as the pulse train progresses, as a change in pulse polarity simply flips the phase but leaves the amplitude unchanged. Thus, since power is proportional to signal level squared, the peak-to-average power ratio (PAPR) of this modulated signal is 1. When the modulating signal is filtered, however, for example, signal $b(t)$ shown in Fig. 4.10, the amplitude of the modulated signal changes with changes in the polarity of the modulating signal, decreasing from maximum to a minimum of 0. The average signal power is thus less than the peak signal power. For limited filtering, the PAPR is likely to be slightly above 1, but for significant filtering, it is more likely to be in the region of 2.

4.3.3   Quadrature Amplitude Modulation (QAM)

The BPSK system described above is only capable of amplitude modulation accompanied by $0°$ or $180°$ phase shifts. However, by adding a *quadrature branch* as shown in Fig. 4.13, it becomes possible to generate signals with any desired amplitude and phase. In the quadrature branch, a second PAM baseband signal is multiplied with a sinusoidal carrier of frequency f_c, identical to that of the *in-phase* carrier but delayed in phase by $90°$. The outputs of the two multipliers are then added together to form a *quadrature amplitude modulated* (QAM) signal.

Labeling the in-phase PAM filtered signal $b_i(t)$ and the quadrature PAM filtered signal $b_q(t)$, then the summed output of the modulator, $c(t)$, is given by

$$c(t) = b_i(t) \cos 2\pi f_c t + b_q(t) \sin 2\pi f_c t \tag{4.23}$$

At the QAM demodulator, the incoming signal is passed through the bandpass filter F_{R1} to limit noise and interference. It is then divided into two and each branch inputted to a multiplier, one multiplier being fed also with an in-phase carrier, $\cos 2\pi f_c t$, and the other with a quadrature carrier, $\sin 2\pi f_c t$.

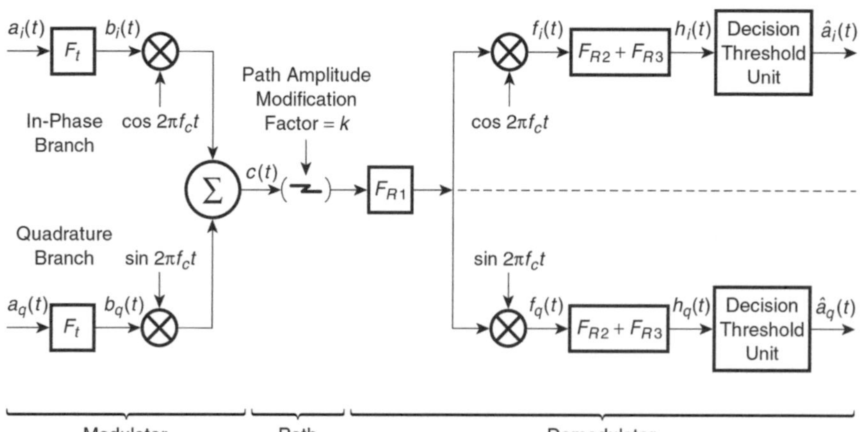

Fig. 4.13 Simplified one-way quadrature amplitude modulated system

The output signal $f_i(t)$ of the in-phase multiplier is given by

$$
\begin{aligned}
f_i(t) &= k \cdot c(t) \cdot \cos 2\pi f_c t \\
&= k \cdot b_i(t) \cos^2 2\pi f_c t + k \cdot b_q(t) \sin 2\pi f_c(t) \cdot \cos 2\pi f_c t \\
&= \frac{k}{2} b_i(t) + \frac{k}{2} b_i(t) \cos\left(2 \cdot 2\pi f_c t\right) + \frac{k}{2} b_q(t) \sin\left(2 \cdot 2\pi f_c t\right)
\end{aligned}
\tag{4.24}
$$

The only difference between $f_i(t)$ and $f(t)$ of Eq. (4.13) for an in-phase only modulated system is the final component of Eq. (4.24). However, this component, like the second component in Eq. (4.24), is spectrally centered at $2f_c$ and is filtered prior to decision threshold detection, leaving only the original quadrature modulating signal $b_i(t)$.

The output signal $f_q(t)$ of the quadrature multiplier is given by

$$
\begin{aligned}
f_q(t) &= k \cdot c(t) \cdot \sin 2\pi f_c t \\
&= k \cdot b_q(t) \sin^2 2\pi f_c t + k \cdot b_i(t) \cos 2\pi f_c t \cdot \sin 2\pi f_c t \\
&= \frac{k}{2} b_q(t) - \frac{k}{2} b_q(t) \cos\left(2 \cdot 2\pi f_c t\right) + \frac{k}{2} b_i(t) \sin\left(2.2\pi f_c t\right)
\end{aligned}
\tag{4.25}
$$

As with the output $f_i(t)$ from the in-phase multiplier, $f_q(t)$ consists of the original in-phase modulating signal $b_q(t)$ as well as two components centered spectrally at $2f_c$ which are filtered prior to decision threshold unit. Thus, by quadrature modulation, it is possible to transmit two independent bitstreams on the same carrier with no interference of one signal with the other, given ideal conditions.

4.3.4   Quadrature Phase-Shift Keying (QPSK)

Quadrature (or quaternary) phase-shift keying (QPSK) is the simplest implementation of quadrature amplitude modulation and is sometimes referred to as *4-QAM*. In it, the modulated signal has four distinct states. A block diagram of a conventional, simplified, QPSK system is shown in Fig. 4.14a. The binary non-return-to-zero (NRZ) input data stream $a_{in}(t)$, of bit rate f_b, and bit duration τ_b, is fed to the modulator where it is converted by a serial to parallel converter into two NRZ streams, an I stream labeled $a_i(t)$ and a Q stream labeled $a_q(t)$, each of symbol rate f_B, half that of f_b, and symbol duration τ_B, twice that of τ_b. The relationship between the data streams $a_{in}(t)$, $a_i(t)$, and $b_q(t)$ is shown in Fig. 4.14b. The I and Q streams undergo standard QAM processing as described in Sect. 4.3.3 above. The in-phase multiplier if fed by the carrier signal $\cos 2\pi f_c t$. The quadrature multiplier is fed by the carrier signal delayed by 90° to create the signal $\sin 2\pi f_c t$. The output of each multiplier is a BPSK signal. The BPSK output signal of the in-phase carrier driven multiplier has phase values of 0° and 180° relative to the in-phase carrier, and the BPSK output signal of the quadrature carrier driven multiplier has phase values of

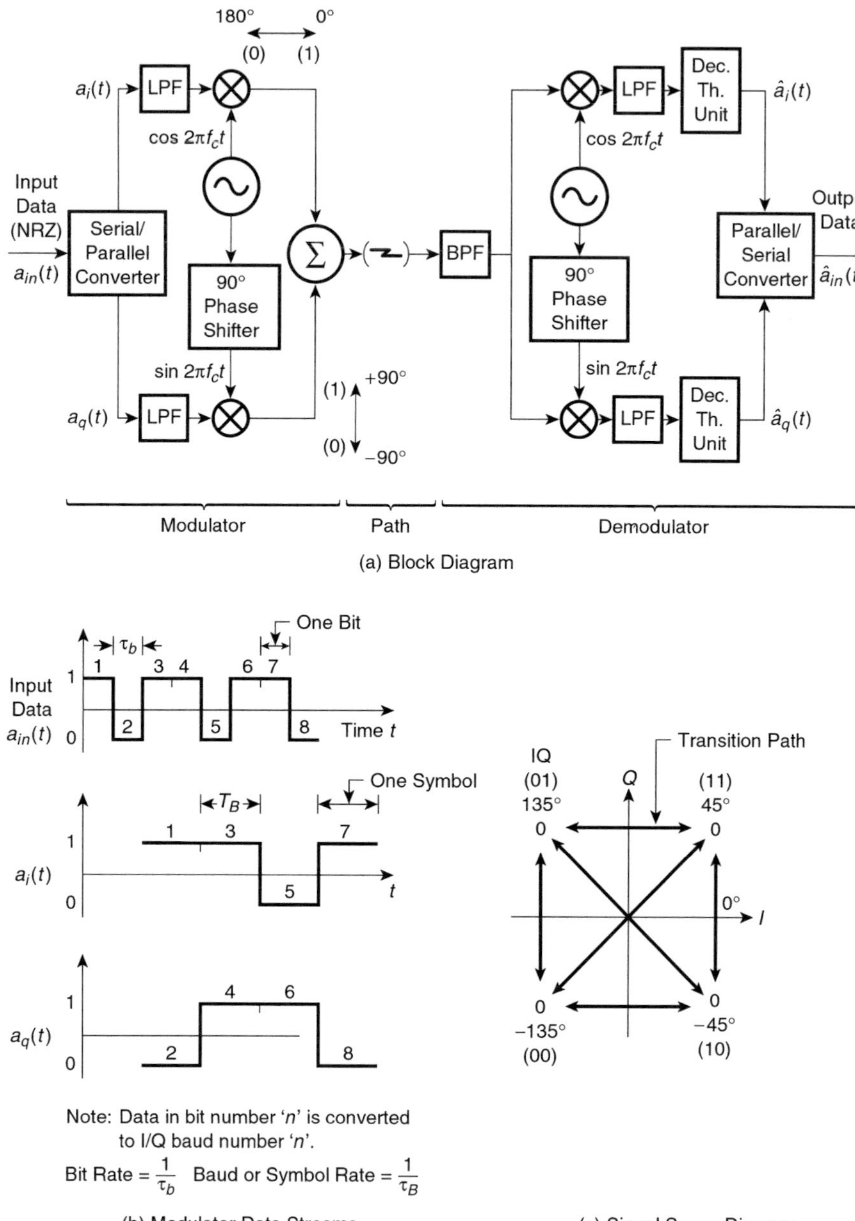

Fig. 4.14 QPSK system representation

90° and 270° relative to the in-phase carrier. The multiplier outputs are summed to give a four-phase signal. Thus, QPSK can be regarded as two *associated BPSK* (ABPSK) systems operating in quadrature.

The four possible output signal states of the modulator, their *IQ* digit combinations, and their possible transitions from one state to another are shown in Fig. 4.14c. We note that either $90°$ or $180°$ phase transitions are possible. As an example, a $90°$ phase transition occurs when the *IQ* combination changes from 00 to 10, and a $180°$ phase transition occurs when the *IQ* combination changes from 00 to 11. For a system where $a_i(t)$ and $a_q(t)$ are unfiltered prior to application to the multipliers, phase transitions occur instantaneously, and thus, the signal has a constant amplitude. However, for systems where these signals are filtered, as is normally the case to limit the radiated spectrum, phase transitions occur over time, and the modulated signal has an amplitude envelope that varies with time. In particular, a $180°$ phase change results in a change over time in amplitude envelope value from maximum to zero and back to maximum.

In the demodulator, as a result of quadrature demodulation, signals $\widehat{a}_i(t)$ and $\widehat{a}_q(t)$, estimates of the original modulating signals, are produced. These signals are then recombined in a parallel to serial converter to form $\widehat{a_{in}}(t)$, an estimate of the original input signal to the modulator.

As indicated above, QPSK can be regarded as two associated BPSK systems operating in quadrature. From a spectral point of view at the modulator output, two BPSK signal spectra are superimposed on each other. The BPSK symbol duration is τ_B. But $\tau_B = 2\tau_b$. Thus, the spectral density of each BPSK signal and hence of the QPSK signal is given by Eq. (4.22), but with τ_b replaced by $2\tau_b$. Making this replacement, we get

$$G_{QPSK}(f) = 2P_S\tau_b \left[\frac{\sin 2\pi(f - f_c)\tau_b}{2\pi(f - f_c)\tau_b} \right] \tag{4.26}$$

A graph of $G_{QPSK}(f)$ is shown in Fig. 4.15. We note that the widths of the main lobe and side lobes are half that for BPSK given the same bit rate for each system. As a result, the maximum spectral efficiency of QPSK is twice that of BPSK, i.e., 2 bits/sec/Hz.

It can be shown [5] that the probability of bit error $P_{be(QPSK)}$ of a QPSK (4-QAM) system, with optimum filtering, and in the presence of white Gaussian noise, is given by

$$P_{be(QPSK)} = Q\left[\left(\frac{2E_b}{N_0} \right)^{\frac{1}{2}} \right] \tag{4.27}$$

We note that this relationship is identical to that for the probability of bit error versus E_b/N_0 for BPSK. Graphs of P_{be} versus E_b/N_0 for QPSK and other linear modulation methods are shown in Fig. 4.16.

In summary, for the same bit rate, the spectral efficiency of QPSK is twice that of BPSK with no loss in the probability of bit error performance in ideal circumstances. The QPSK hardware is, however, more complex than that required for BPSK.

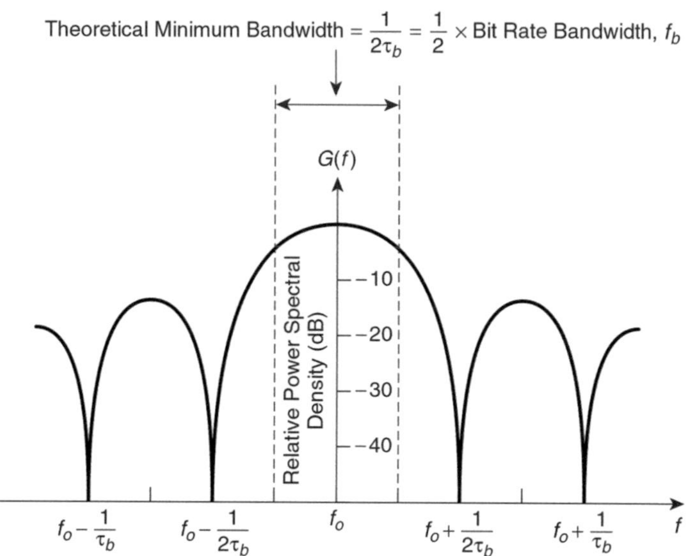

Fig. 4.15 Power spectral density of QPSK

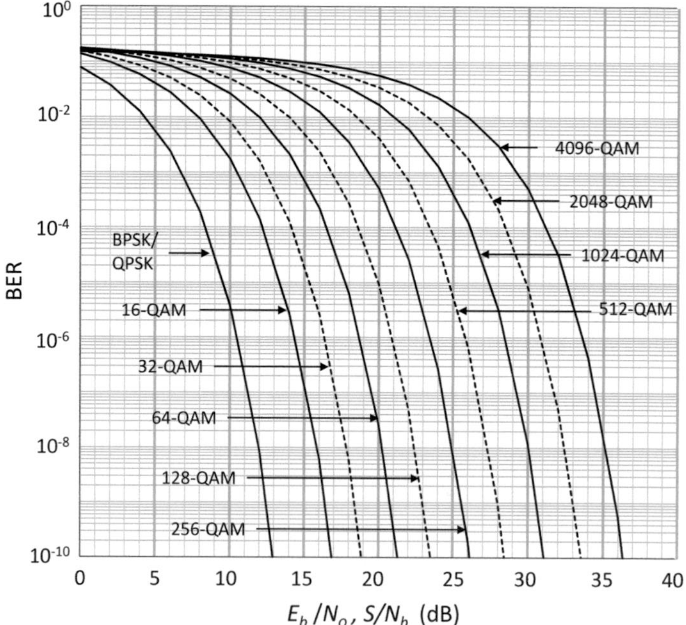

Fig. 4.16 BER versus E_b/N_0 (S/N_b) for linear modulation methods

Further, in transmission through nonlinear components such as power amplifiers, filtered QPSK is subject to quadrature crosstalk, a situation where modulation on one quadrature channel ends up on the other. This situation can also arise in the receiver if the phase difference between the coherent detection oscillators is not kept to 90°. Thus, in real-world environments, BPSK is a more robust modulation scheme than QPSK.

4.3.5   High-Order 2^{2n}-QAM

Though relatively easy to implement and robust in performance, linear four phase systems such as QPSK do not often afford the desired spectral efficiency in commercial wireless systems. Higher-order QAM systems, however, do permit higher spectral efficiencies and have become very popular. A common class of QAM systems allowing high spectral density is one where the number of states is 2^{2n}, where n equals 2,3, 4,.... A generalized and simplified block diagram of a 2^{2n}-QAM system is shown in Fig. 4.17. The difference between this generalized system and the QPSK system shown in Fig. 4.14 is that (a) in the generalized modulator, the I and Q signals $a_i(t)$ and $a_q(t)$ are each fed to a 2 to 2^n level converter prior to filtering and multiplication with the carrier and (b) in the generalized demodulator, the outputs of the decision threshold units are each fed to a 2^n to 2 level converter prior to being combined in a parallel to serial converter. 2^{2n}-QAM wireless systems have been deployed commercially for values of n from 2 to 7.

For n equal 2, a 16-QAM system is derived. In such a system, incoming symbols to each modulator level converter are paired, and output symbols, in the form of signals at one of four possible amplitude levels, are generated in accordance with the coding table shown in Fig. 4.18a. The duration of these output symbols, τ_{B4L} say, is twice that of τ_B, the duration of the level converter input symbols. As a result of the application of the four-level signals to the multipliers, the output of each multiplier is a four-level amplitude-modulated DSBSC signal, and the combined signal at the

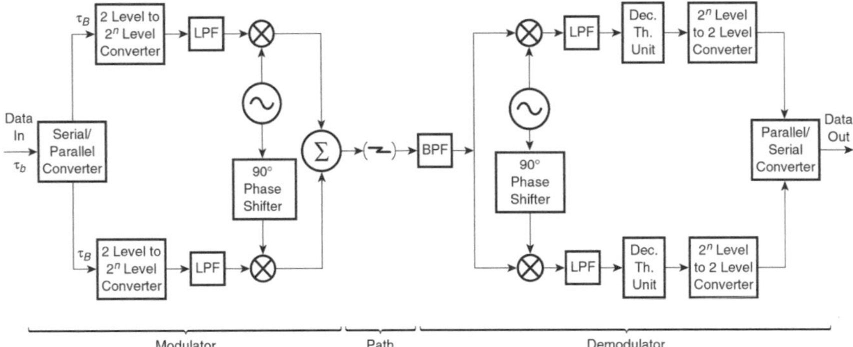

Fig. 4.17 Generalized block diagram of a 2^{2n}-QAM system

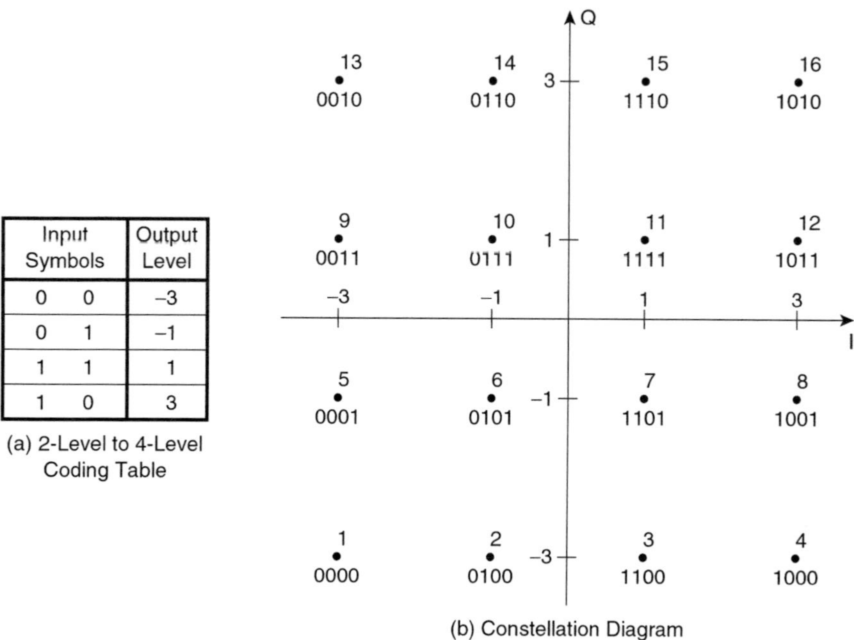

(a) 2-Level to 4-Level
Coding Table

(b) Constellation Diagram

Fig. 4.18 16-QAM level converter coding table and constellation diagram

modulator output is a QAM signal with 16 states. Thus, 16-QAM can be treated as two four-level PAM DSBSC systems operating in quadrature. The constellation diagram of a 16-QAM signal is shown in Fig. 4.18b. From this figure, it is clear that 16-QAM has an amplitude envelope that varies considerably over time, irrespective of whether the signal has been filtered or not, and thus must be transmitted over a highly linear system if it is to preserve its spectral properties.

As τ_B, the duration of symbols from the serial to parallel converter, is equal to $2\tau_b$, where τ_b is the duration of incoming bits to the modulator, it follows that

$$\tau_{B4L} = 4\tau_b \tag{4.28}$$

Using Eq. (4.28) and the same logic used to determine the spectral density of QPSK, it can be shown that $G_{16-QAM}(f)$, the spectral density of 16-QAM, is given by

$$G_{16-QAM}(f) = 4P_s\tau_b \left[\frac{\sin 4\pi(f - f_c)\tau_b}{4\pi(f - f_c)\tau_b} \right] \tag{4.29}$$

$G_{16-QAM}(f)$ is such that its main lobes and side lobes are one-fourth as wide as those of BPSK. As a result, the maximum spectral efficiency off 16-QAM is 4 bits/sec/Hz, twice that of QPSK.

The two-level to four-level coding shown in Fig. 4.18a is an example of *Gray coding*. In Gray coding, the bits that create any pair of adjacent levels differ by only one bit. It is interesting to note that, as a result of Gray coding in the I and Q channels, the 16 states shown in the signal space diagram in Fig. 4.18b are also Gray coded. Thus, an error resulting from one of these states being decoded as one of its closest adjacent states will result in only one bit being in error. Since there are four bits in a symbol, the bit error rate will be only one-fourth of the symbol error rate.

For a 16-QAM system, it can be shown [5] that the probability of bit error P_{be} $_{(16-QAM)}$ when Gray coded is given by

$$P_{be(16-QAM)} = \frac{3}{4}Q\left[\left(\frac{4}{5}\frac{E_b}{N_0}\right)^{\frac{1}{2}}\right] \tag{4.30}$$

A graph of $P_{be(16-QAM)}$ versus E_b/N_0 is shown in Fig. 4.16. It will be observed that for a probability of bit error of 10^{-3} the E_b/N_0 required for 16-QAM is 3.8 dB greater than that required for QPSK. Thus, the doubling of the spectral efficiency achieved by 16-QAM relative to QPSK comes at the expense of probability of bit error performance.

For n equal 3, a 64-QAM system is derived, two eight-level PAM DSBSC signals being combined in quadrature. The eight-level PAM signals are created by grouping the incoming symbols to the level converter into sets of three and using these three-digit code words to derive the 8 output levels.

Using the same logic as applied to the above analysis of 16-QAM but with generalized equations, it can be shown that the maximum spectral efficiency of 2^{2n}-QAM is given by

$$\eta_{2^{2n}-QAM} = 2n\,\text{bits}/\sec/\text{Hz} \tag{4.31}$$

Thus, that of 64-QAM is 6 bits/sec/Hz, and that of 4096-QAM is 12 bits/sec/Hz.

It is shown in [5] that for Gray coded 2^{2n}-QAM, with optimum filtering, and in the presence of white Gaussian noise, the generalized equation for the probability of bit error P_e versus E_b/N_0 is

$$P_{be(2^{2n}-QAM)} = \frac{2}{\log_2 L}\left[1-\frac{1}{L}\right]Q\left[\left(\frac{6\log_2 L}{L^2-1}\frac{E_b}{N_0}\right)^{\frac{1}{2}}\right] \tag{4.32}$$

where $L = 2^n$.

Thus, for 64-QAM, we have

$$P_{be(64-QAM)} = \frac{7}{12}Q\left[\left(\frac{2}{7}\frac{E_b}{N_0}\right)^{\frac{1}{2}}\right] \tag{4.33}$$

for 256-QAM, we have

$$P_{be(256-QAM)} = \frac{15}{32} Q\left[\left(\frac{8}{85}\frac{E_b}{N_0}\right)^{\frac{1}{2}}\right] \tag{4.34}$$

for 1024-QAM, we have

$$P_{be(1024-QAM)} = \frac{31}{80} Q\left[\left(\frac{10}{341}\frac{E_b}{N_0}\right)^{\frac{1}{2}}\right] \tag{4.35}$$

and for 4096-QAM, we have

$$P_{be(4096-QAM)} = \frac{21}{64} Q\left[\left(\frac{4}{455}\frac{E_b}{N_0}\right)^{\frac{1}{2}}\right] \tag{4.36}$$

The probability of bit error relationships for 4, 16, 64, 256, 1024, and 4096-QAM are shown as unbroken lines Fig. 4.16. We observe from Fig. 4.16 that as the number of QAM states increases, the P_e performance decreases, greater and greater E_b/N_0 and hence signal-to-noise ratio (see Eq. 4.20) being required for the same bit error rate.

In the QAM realizations discussed above, the I and Q carriers were modulated via first a parallel-to-series converter, followed by, for the cases where n was greater than 1, a 2-level to 2^n level converter. We note, however, that though helpful in conveying the modulation conceptually, such a physical realization is not necessary. All that is necessary is to utilize any mapping structure that converts each grouping of 2^n incoming data bits to the desired I and Q modulating values. Thus, with 16-QAM for example, the mapper needs only be programmed to take any incoming 4-bit combination and map it to the I and Q values shown in Fig. 4.18. For example, incoming bits 1110 are mapped to an I value of 1 and a Q value of 3. Viewed another way, it's mapped to the complex value $1 + j3$.

4.3.6   High-Order 2^{2n+1}-QAM

For the $2^{2n} - QAM$ systems described above, the signal space consists of a full rectangular array of states with each state coded by an even number of bits. When each state is coded by an odd number of bits, however, it's not possible to create a full rectangular array. Such systems have 2^{2n+1} states with each state coded with $2n + 1$ bits. For n equals 1, we get an 8-QAM system. Many constellations have been proposed for 8-QAM. However, as there appears to be no obvious "standard" constellation and as 8-QAM is rarely used in fixed wireless communications, it

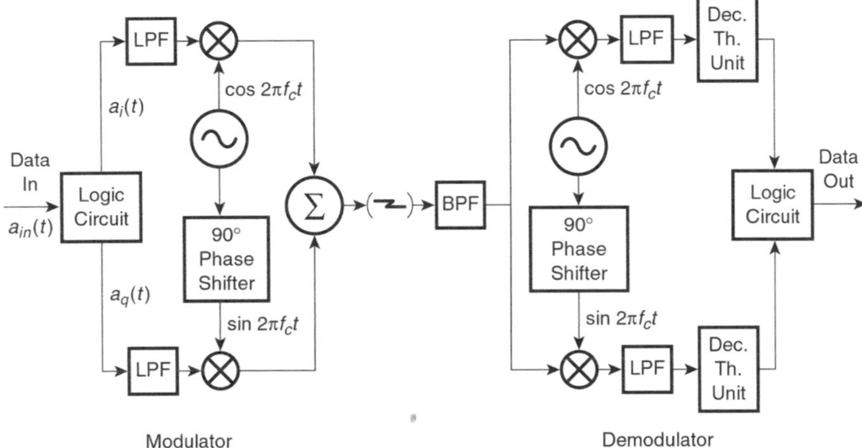

Note: The Modulator Logic Circuit uses a look up table to convert odd number input sequences to I and Q symbol streams. The Demodulator Logic Circuit uses a look up table to convert I and Q symbol streams to odd number output sequences.

Fig. 4.19 Simplified block diagram of 2^{2n+1}-QAM system

won't be reviewed here other than to point out that it affords a spectral efficiency of 3 bits/sec/Hz. For $n = 2, 3, 4, 5$ and 6, we get 32-QAM, 128-QAM, 512-QAM, 2048-QAM, and 8192-QAM, respectively. 2^{2n+1}-QAM wireless systems have been deployed commercially for values of n from 2 to 6.

With 2^{2n+1}-QAM modulators, because an odd number of bits is used to code each signal state, it is not possible to divide the incoming bitstream into two and then use two-level to L-level PAM modulation in order to create the QAM signal. Instead, the incoming signal of bit rate f_b and duration τ_b is fed to a logic circuit that uses a look up table to create in-phase and quadrature L level symbols of duration τ_B, where

$$\tau_B = (2n + 1)\tau_b \tag{4.37}$$

In the demodulator, the L level outputs of the I and Q decision threshold units are fed to a logic circuit which, via its lookup table, recreates the original bits. A simplified block diagram of a 2^{2n+1}-QAM system is shown in Fig. 4.19.

Because of the freedom afforded by the lookup table design, many 2^{2n+1} signal state constellations are possible. A very common class is the *cross-constellation* class, so-called because of their cross appearance. Constellations in this class exist where n is ≥ 2 and are essentially full rectangular $L \times L$ arrays, where $L = 3 \cdot 2^{n-1}$, but with corner states removed.

Figure 4.20 shows a 32-QAM cross-constellation. It is a 6×6 array with the four corner states removed. From the figure, it is clear that this constellation can be created by summing two six level PAM DSBSC signals in quadrature. Each state in the 32-QAM constellation is equally likely. However, unlike 2^{2n}-QAM, each level in each of the six-level DSBSC signals is not equally likely. In Fig. 4.20, the

Fig. 4.20 32-QAM cross-constellation

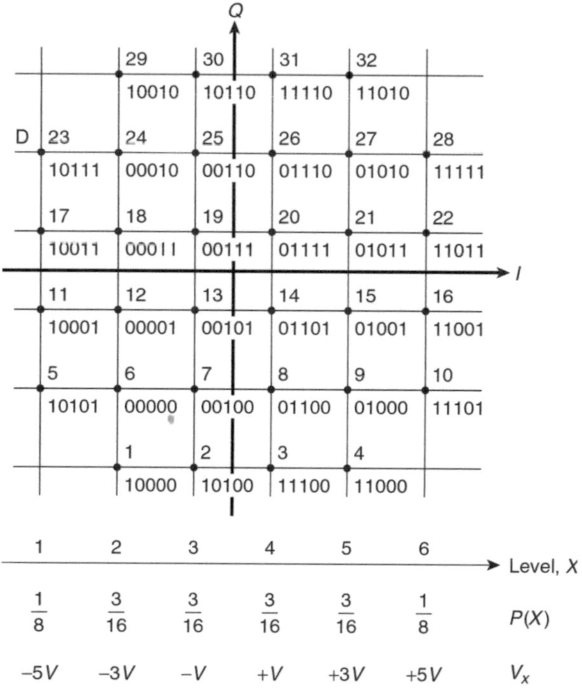

probability of level X occurrence, $P(X)$, in the in-phase channel is shown. Also shown in Fig. 4.20 are the six PAM levels that drive the in-phase multiplier. The same probabilities and PAM levels apply to the equivalent levels in the quadrature channel. We note that the constellation is not fully Gray coded. Specifically, states 5, 10, 23, and 28 each differ by more than one bit for their closest neighbors. Thus, an error resulting from one of these four states being decoded as one of its closest adjacent states will result in more than one bit being in error. Since there are five bits in a symbol, the bit error rate will be greater than one-fifth of the symbol error rate. However, as most of the symbols don't differ by more than one bit from their closest neighbors, the actual bit error can be assumed to be for practical purposes that achieved by full Gray coding, that is, one-fifth of the symbol error rate.

Figure 4.21 shows typical 32-QAM modulator logic circuit signals. We note that the first 5 bits of data in are 10110. These bits create constellation point 30, which is at position $-V$ on the in-phase axis and $+5$ V on the quadrature axis. The signals $a_i(t)$ and $a_q(t)$ shown in Fig. 4.19 are thus as shown in Fig. 4.21.

Using the same logic as applied to the above analysis of 16-QAM, but with generalized equations, it can be shown that the maximum spectral efficiency of 2^{2n+1}-QAM systems is given by:

$$\eta_{2^{2n+1}-QAM} = 2n + 1 \text{ bits/ sec /Hz} \tag{4.38}$$

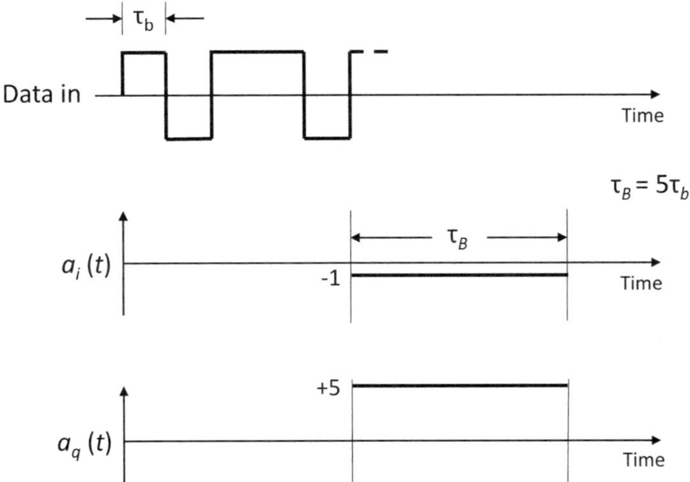

Fig. 4.21 32-QAM modulator logic circuit signals

Thus, that of 32-QAM is 5 bits/sec/Hz, and that of 2048-QAM is 11 bits/sec/Hz.

Determining the probability of error of cross-constellation 2^{2n+1}-QAM systems is not as straightforward as for square constellation 2^{2n}-QAM systems. However, a simple and elegant (though tedious for large constellations) process of accomplishing this is given by Burr in Sect. 2.5 of [6] and is detailed in Appendix E. Following are the bit error rates derived using this method for 32, 128, 512, and 2048-QAM using this method, and assuming Gray coding, even though, as discussed above, this is not fully possible for cross-constellations:

$$P_{be(32-QAM)} = \frac{13}{20} Q\left(\sqrt{\frac{E_b}{2N_0}}\right) \tag{4.39}$$

$$P_{be(128-QAM)} = \frac{29}{56} Q\left(\sqrt{\frac{7E_b}{41N_0}}\right) \tag{4.40}$$

$$P_{be(512-QAM)} = \frac{61}{144} Q\left(\sqrt{\frac{9E_b}{165N_0}}\right) \tag{4.41}$$

$$P_{be(2048-QAM)} = \frac{31}{88} Q\left(\sqrt{\frac{11E_b}{661N_0}}\right) \tag{4.42}$$

These probability of bit error rate relationships are shown as dashed lines in Fig. 4.16.

4.3.7   Peak-to-Average Power Ratio

For the unfiltered BPSK, QPSK, and QAM systems studied above, the signal level is constant during each entire symbol. Thus, for BPSK and QPSK, the symbol level is constant throughout an entire symbol stream, and as a result, the peak level equals the average level, and peak power equals average power. Therefore, the peak-to-average power ratio (PAPR) is 1 or 0 dB. For 2^{2n}-QAM and 2^{2n+1}-QAM systems, however, where $n > 1$, the symbols can take several levels, leading to a PAPR greater than 1. Consider the 16-QAM constellation shown in Fig. 4.22. Here, we see that the symbols can take one of four different levels. The highest level is $\sqrt{18}x$, and thus peak power is $18x^2$. The average power, $\overline{X^2}$, is given by

$$\overline{X^2} = \frac{4 \times 2x^2 + 8 \times 10x^2 + 4 \times 18x^2}{16} = 10x^2 \tag{4.43}$$

Thus, for unfiltered symbols, the PAPR of 16-QAM equals $18x^2/10x^2 = 1.8 = 2.55$ dB. It can similarly be shown that, for unfiltered symbols, the PAPR of 64-QAM is 3.69 dB, and that of 256-QAM is 4.23 dB.

Shown in Table 4.1 is the PAPR for rectangular constellation 2^{2n}- QAMs from 4-QAM through 4096-QAM for unfiltered symbols. Shown in Table 4.2, the PAPR

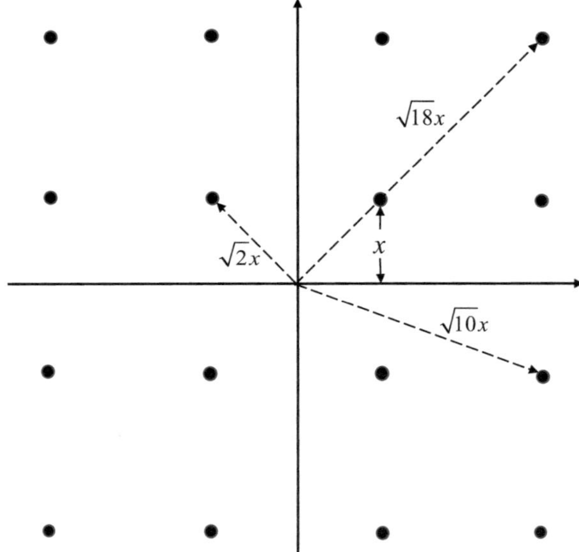

Fig. 4.22 Symbol levels on a 16-QAM constellation diagram

Table 4.1 PAPR for rectangular constellation 2^{2n}- QAMs

	PAPR (dB)				
4-QAM	16-QAM	64-QAM	256-QAM	1024-QAM	4096-QAM
0	2.55	3.69	4.23	4.50	4.66

Table 4.2 PAPR for cross-constellation 2^{2n+1}- QAMs

	PAPR (dB)		
32-QAM	128-QAM	512-QAM	2048-QAM
2.30	3.17	3.59	3.80

for cross-constellation 2^{2n+1}- QAMs from 32-QAM through 2048-QAM for unfiltered symbols.

We note that for cross-constellation 2^{2n+1}-QAM systems, the peak-to-average power ratio for unfiltered symbols is, relatively speaking, better than for rectangular 2^{2n}-QAM systems as the corners of the $L \times L$ forming arrays are removed. For example, the PAPR of 32-QAM is 2.30 dB, less than of 16-QAM which is 2.55 dB.

It is also possible to modify the square constellation structure of 2^{2n}-QAM systems such that the four high amplitude corners are removed and the PAPR reduced. For 64-QAM and 256-QAM, such a PAPR reduction can be achieved should the standard signal point square constellation diagram be modified to be non-square as shown by Morais in [7]. For 64-QAM, where the PAPR of the square constellation is 3.7 dB, the proposed constellation, which has the general shape shown in Fig. 4.23, exhibits a PAPR of 2.50 dB, for a net reduction of 1.2 dB and, hence, for a given power amplifier, an increase in maximum output power of 1.2 dB. For 256-QAM, where the PAPR of the square constellation is 4.2 dB, the proposed constellation exhibits a PAPR of 3.0 dB, again for a net reduction also of 1.2 dB and, hence, for a given power amplifier, an increase in maximum output power also of 1.2 dB.

For these non-square constellations, the bit error rate performances are slightly degraded relative to their associated square constellations. For the non-square 64-QAM constellation, the increase in required SNR for a BER of between 10^{-3} and 10^{-6} is approximately 0.3 dB. Thus, the net link margin improvement with its use would be approximately $1.2–0.3 = 0.9$ dB. For the non-square 256-QAM constellation, the increase in required SNR for a BER of between 10^{-3} and 10^{-6} is approximately 0.4 dB. Thus, the net link margin improvement with its use would be approximately $1.2–0.4 = 0.8$ dB. A method to perform hard and soft bit demapping for these non-square constellations is given in [8].

In the discussion above, we have studied unfiltered symbols where PAPR analysis is straightforward. In practical QAM systems, however, symbols are always filtered so as to reduce occupied bandwidth. When symbols are filtered, the analysis of the resulting PAPR becomes much more complex. Such analysis shows that the unfortunate effect of this filtering is to significantly increase the PAPR. To understand why, consider unfiltered symbols. Here, the symbol is at full amplitude for the entire symbol period. Thus, when averaged over a large number of symbols, the full amplitude per symbol is accounted for. When the symbols are filtered, however, each symbol is only at full amplitude at the center of its period. Elsewhere, its amplitude is less than the peak, rising towards the peak and descending after the peak. Further, the more the filtering, i.e., the less the excess bandwidth, the less is the integrated area of the symbol over the symbol period. The net result is that the average symbol power is reduced relative to unfiltered symbols, this average

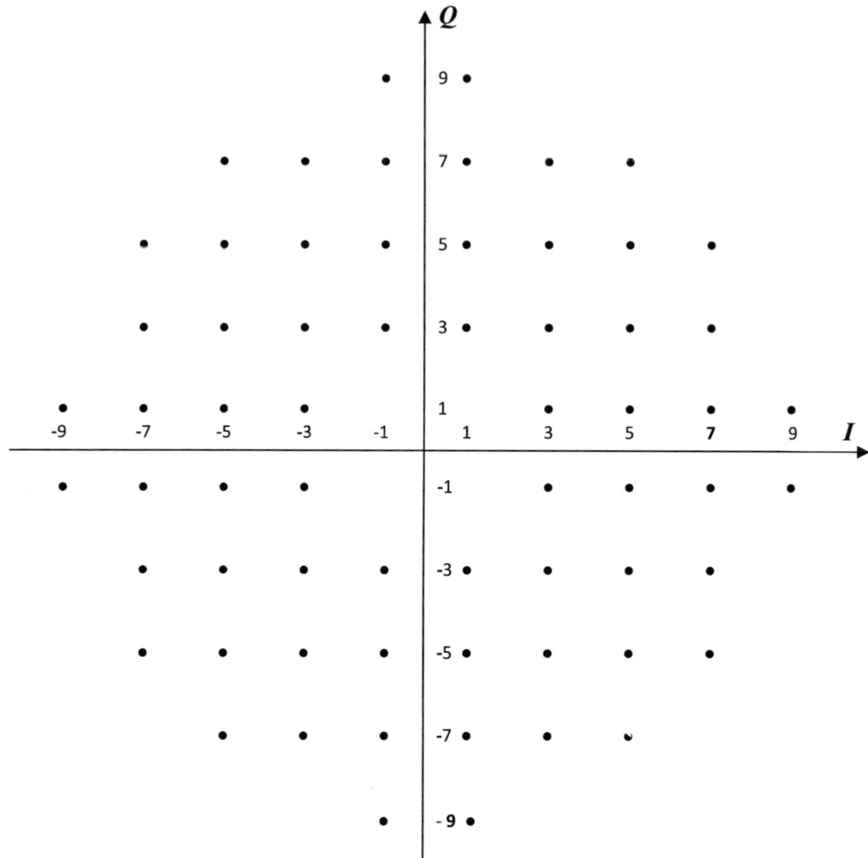

Fig. 4.23 Non-square 64-QAM signal point constellation (Word)

power reducing as filtering increases. Peak power, however, which occurs at the center of a symbol's period, is unaffected by filtering. The overall effect is an increase in PAPR due to filtering that increases as filtering increases. This increase in PAPR does not, however, change the relative PAPR performance between the various constellations.

4.4   Transmission IF and RF Components

For the modems described above, if the local oscillator frequency in the transmitter and that in the receiver is set to the actual RF transmission center frequency, then we get what's called *direct conversion* (*homodyne conversion*). Here, in the transmitter, the baseband signal is directly converted to the system RF bandwidth, and in the receiver, the received signal is directly converted to baseband. However, often, these

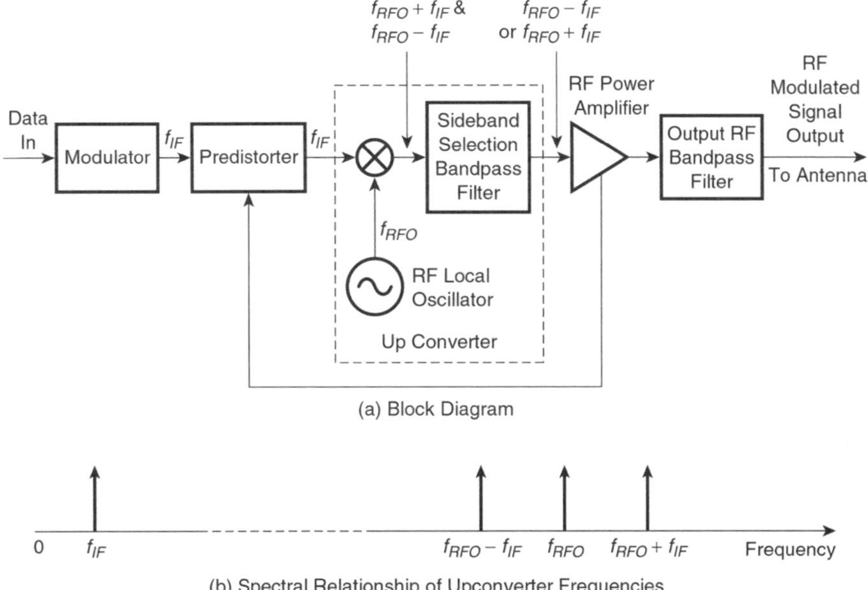

(a) Block Diagram

(b) Spectral Relationship of Upconverter Frequencies

Fig. 4.24   Digital heterodyne transmitter

modems do not operate at the frequency of actual RF transmission but, instead, at a lower frequency. The modem frequency is then referred to as the *intermediate frequency* (IF). There are many advantages to the modem being at an IF frequency. Among these are a standard design which greatly aids production, efficiency, design at a frequency where components are cost effective, and design that allows complex signal processing such as high-order bandpass filtering. To effect transmission at the desired RF frequency, the heterodyne principle, i.e., up and down signal conversion, is employed, and this approach is therefore referred to as *heterodyne conversion*. To effect this, additional components are added to the modulator and demodulator to create the full transmitter and receiver. Figure 4.24 shows a digital transmitter with these components, and Fig. 4.25 shows the associated receiver. In this section, we discuss these components.

4.4.1   Transmitter Upconverter and Receiver Downconverter

The *transmitter upconverter*, shown in Fig. 4.24a, translates the modulator output IF signal of frequency f_{IF} up to the desired RF frequency. It accomplishes this by mixing (multiplying) the IF signal with a RF local oscillator signal. If the IF signal is $m(t) \cos (2\pi f_{IF}t)$, where $m(t)$ represents the modulation on the carrier, and the RF local oscillator signal is $\cos(2\pi f_{RFO}t)$, then the output of the mixer $s_{RF}(t)$ is given by

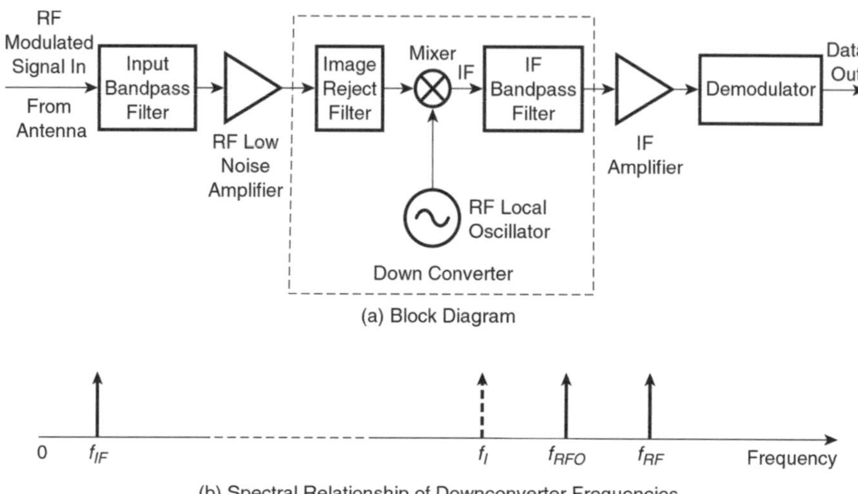

Fig. 4.25 Digital heterodyne receiver

$$s_{RF}(t) = m(t)\cos(2\pi f_{IF}t)\cdot\cos(2\pi f_{RFO}t)$$
$$= \frac{1}{2}m(t)\cos(2\pi(f_{RFO}+f_{IF})t) + \frac{1}{2}m(t)\cos(2\pi(f_{RFO}-f_{IF})t) \qquad (4.44)$$

Thus, the output of the mixer consists of two sideband signals, one below and one above the RF oscillator frequency by amounts equal to the IF frequency. The spectral relationship of the frequencies associated with *upconversion* is shown in Fig. 4.24b. The designer chooses f_{RFO} so that one of these sideband signals is the desired one. The undesired sideband is eliminated by passing the output of the mixer through a bandpass filter centered on the desired signal.

The *receiver downconverter*, shown if Fig. 4.25a, does the opposite of the upconverter. It translates, via *downconversion*, the received modulated RF signal, of center frequency f_R, down to the IF frequency f_{IF}. Note, however, that if the desired received frequency is above the RF local oscillator frequency, then the process also translates to IF any signal at a frequency below that of the RF local oscillator that's offset by the IF frequency and vice-versa. This undesired frequency is called the *image frequency, f_I* say, and any received interfering signal at or close to this frequency must be eliminated by a filter placed ahead of the mixer. Note, also, that even if an interfering signal is not present at the image frequency, there is always thermal noise there that, if not removed, is downconverted and results in a doubling of the noise at the demodulator input. In Fig. 4.25a, an image reject filter is shown just ahead of the mixer. This filter, however, is only necessary if the bandwidth of the input filter is such as to not reject the image frequency. The spectral relationship of the frequencies associated with downconversion is shown in Fig. 4.25b.

Some receivers, for frequency agility purposes or to allow the sharing of one RF local oscillator between the receiver and the accompanying transmitter, employ a double downconversion process and hence have two IF frequencies.

4.4.2   Transmitter RF Power Amplifier and Output Bandpass Filter

The transmitter RF *power amplifier* follows the upconverter, and its purpose is to provide a high level of output power so that an adequate signal level is available to the receiver even with significant fading. The output power of fixed wireless transmitters varies from slightly less than a tenth of a Watt to several Watts, with maximum attainable power decreasing with frequency.

For linear modulation systems such as QAM ones with signal states of varying amplitudes, linear amplification is essential to maintain acceptable performance, with higher and higher linearity required as the number of modulation states increases. The effects of nonlinearity on such systems are a nonlinear displacement of signaling states in the phase plane and the regeneration of spectral side lobes removed by prior filtering. The displacement of the signaling states degrades the error probability performance, while the regenerated spectrum can cause interference to signals in adjacent channels. To avoid these effects, the power amplifier must be capable of linearly amplifying all signaling states and thus amplifying the peak signal power. However, high-order linear modulation results in signals where the ratio between peak power and average power can be several dBs. For example, for filtered 16-QAM, 64-QAM, and 256-QAM, this ratio is on the order of 6 to 7 dB, 7 to 8 dB, and 8 to 9 dB, respectively. Thus, power amplifiers processing these signals must operate at an average power that is backed off from the peak linear power available by a minimum of the peak to average ratio of the amplified signal. The predistorter shown in Fig. 4.24a improves peak linear power (Sect. 5.6).

The output signal of the RF power amplifier is normally fed via a bandpass filter to the antenna. The bandwidth of this filter is typically wider than the signal spectrum, and its purpose is to allow filtering of out-of-band signals and duplexing of the in-band signal with an associated incoming signal.

4.4.3   The Receiver "Front End"

In a heterodyne receiver, the term *front end* is normally used to describe the receiver input bandpass filter, the low noise amplifier, the downconverter, and the succeeding IF amplifier. In a direct conversion receiver, it normally describes the input bandpass filter and the low noise amplifier. The purposes of the input filter are to eliminate unwanted signals and, in the case of the heterodyne receiver, if no image reject filter is equipped just ahead of the mixer, to filter out frequencies at or close to the image frequency. Following the input filter, there is normally a low noise amplifier, which plays a large part in determining the overall noise performance of the receiver. The characteristic of the receiver that determines the signal-to-noise ratio presented to the coherent demodulator input is the receiver *noise figure, F*. For the heterodyne receiver shown in Fig. 4.25a, the noise figure describes the deterioration of the signal-to-noise ratio from the receiver RF input to the IF amplifier output and hence the demodulator input, due to the presence of all the circuitry between this input and output. For the direct conversion receiver, it describes the deterioration of the signal-to-noise ratio from the receiver RF input to the low noise amplifier output and hence the demodulator input. It is given by:

$$F = \frac{P_{RFSi}/P_{RFni}}{P_{DSi}/P_{Dni}} \tag{4.45}$$

where

$P_{RFSi} = $ RF input signal power,
$P_{DSi} = $ Demodulator input signal power
$P_{RFni} = $ RF input noise power in a frequency band *df*
$P_{Dni} = $ Demodulator input noise power in a frequency band *df*.

Thus

$$P_{DSi}/P_{Dni}(dB) = P_{RFSi}(dBm) - P_{RFni}(dBm) - F(dB) \tag{4.46}$$

where 1 dBm = 1 milliWatt.

The signal-to-noise ratio at the demodulator input, P_{DSi}/P_{Dni}, is that which determines the error rate performance as a result of the input noise. Thus, Eq. (4.46) is important because it indicates that P_{DSi}/P_{Dni} can be determined from a knowledge of the signal-to-noise ratio at the receiver input and the receiver noise figure.

The thermal noise power in Watts available in a small frequency band *df* Hertz from a source having a noise temperature T degrees Kelvin is given by

$$P_n = k \cdot T \cdot df \tag{4.47}$$

where $k = 1.38 \cdot 10^{-23}$ Joules/degree Kelvin (Boltzmann's constant).

For a terrestrial wireless system, the source of thermal noise at the receiver input is the receiving antenna. Antenna noise temperature is normally assumed to be 290° Kelvin. At this temperature, the antenna noise transferred to the receiver is given by Eq. (4.47) to be -174 dBm per Hertz of bandwidth. Thus, the input thermal noise in the bit rate bandwidth f_b Hertz of a digital wireless system is given by

$$P_{RFnib}(dBm) = -174 + 10 \log_{10} f_b \tag{4.48}$$

Substituting Eq. (4.48) into Eq. (4.46) gives the ratio of the demodulator input signal power P_{DSi} to demodulator input noise power in the bit rate bandwidth P_{Dnib} to be

$$P_{DSi}/P_{Dnib}(dB) = P_{RFSi}(dBm) + 174 - 10\log_{10}f_b - F(dB) \qquad (4.49)$$

Recognizing that

$$\frac{P_{DSi}}{P_{Dnib}} = \frac{E_b f_b}{N_0 f_b} = \frac{E_b}{N_0} \qquad (4.50)$$

where E_b is the energy per bit at the demodulator input,
and N_0 is the noise power spectral density at the demodulator input.

Then, Eq. (4.49) can be restated as

$$E_b/N_0(dB) = P_{RFSi}(dBm) + 174 - 10\log_{10}f_b - F(dB) \qquad (4.51)$$

Thus, knowing the received input signal level, the bit rate, and the receiver noise figure, one can calculate the E_b/N_0 at the demodulator input and, from the appropriate probability of error versus E_b/N_0 relationship, the theoretical probability of error.

4.5   Modem Realization Techniques

The modems described above provide a theoretical understanding of their operation and performance. However, to realize such modems in practice, several techniques must be applied. For example, in the demodulator, the identical carrier as available in the modulator has been assumed, but how in practice is this made available? In this section, a number of key implementation techniques, required even for performance in a linear transmission environment, are reviewed. Figure 4.26a shows a quadrature-type modulator that indicates the placement of the techniques covered. Similarly, Figure 4.26b indicates the placement in the associated demodulator.

4.5.1   *Scrambling/Descrambling*

Following the input data to the digital modulator, there is normally a scrambler. Its function is to eliminate from the incoming data stream (a) any periodic data pattern, (b) long sequences of ones or zeros in the incoming signal, and (c) any direct current (DC) component that may occur as a result of these long sequences. Scrambling is achieved, as shown in Fig. 4.27, by generating a repetitive but long pseudorandom bit sequence and logically combining the generated sequence with incoming data.

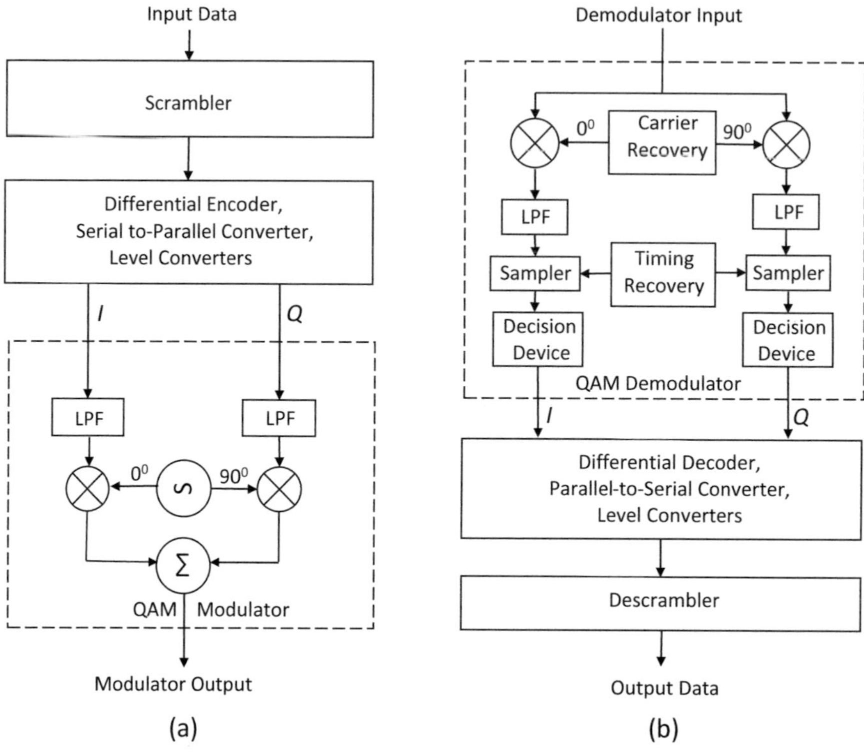

Fig. 4.26 Modem realization techniques

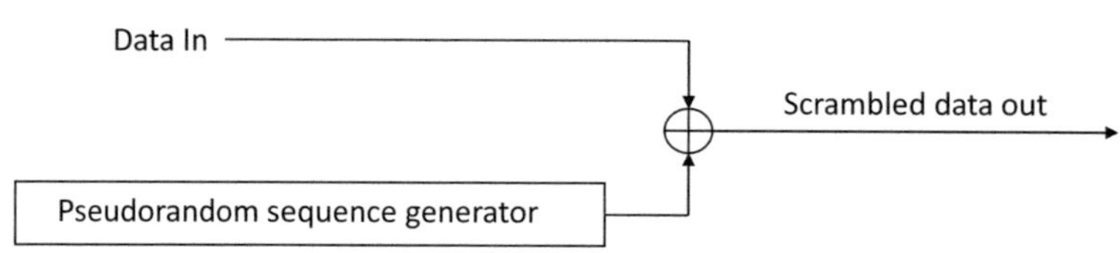

Fig. 4.27 Scrambler

The scrambled output assumes properties similar to the pseudorandom sequence, irrespective of input data properties.

The removal of any DC component by the scrambler allows the use of alternating current (AC) coupled circuitry which in normally easier to implement. By eliminating periodic patterns in the modulating data, scrambling guarantees that the radiated spectrum is essentially uniformly distributed and free of spectral lines. Such lines, if significant, can cause unwanted interference in adjacent RF channels. In the demodulator, the frequent transitions in the demodulated symbols resulting from the elimination of long sequences of ones or zeros are required for accurate timing recovery.

The output of the demodulator parallel to serial converter is fed to a descrambler in order to generate the estimate of the original data stream. Because the descrambler has to know the specifics of the pseudorandom generator in the scrambler in order to function, the scrambler/descrambler circuitry serves as a form of encryption. In fact, some manufacturers provide user programmable scrambler/descrambler circuits so as to give the user direct control over the scrambling sequence.

4.5.2 Carrier Recovery

In digital transmission, the chain includes multiple oscillators. In the transmitter, there is the modulator oscillator. If there is direct conversion to RF by the modulator, then this is the only oscillator, but if the modulator produces an intermediate frequency (IF), there is upconversion following the modulator to RF, and thus, there is a second oscillator. In the receiver, if there is direct conversion to baseband by the demodulator, then the demodulator oscillator is the only oscillator, but if there is downconversion to an intermediate frequency, then there is a second oscillator. For effective recovery of a transmitted QAM signal in the receiver, demodulation must be carried out by an oscillator that is essentially identical in frequency and phase to the incoming signal.

A constant frequency error between the received carrier and the local oscillator corresponds to a continuous time varying rotation of the QAM constellation. Thus, frequency synchronization to remove this rotation spinning is essential to commence effective demodulation. If frequency synchronization is not achieved, then instead of demodulation bringing the modulating signal to baseband, the signal will be near the baseband and hence highly corrupted. However, frequency synchronization alone is not sufficient. A constant carrier phase error corresponds to a fixed rotation of the constellation which has performance degrading effects. Consider a QAM signal given by

$$s_{QAM}(t) = a(t) \cos 2\pi f_c t - b(t) \sin 2\pi f_c t \qquad (4.52)$$

Assume now that this signal is demodulated by the two locally generated quadrature carriers

$$c_i(t) = \cos\left(2\pi f_c t - \theta\right) \qquad (4.53)$$

and

$$c_q(t) = -\sin\left(2\pi f_c t - \theta\right) \qquad (4.54)$$

Multiplying $s_{QAM}(t)$ by $c_i(t)$ followed by a low-pass filter generates the in-phase component

$$y_i(t) = \frac{1}{2}a(t)\cos\theta - \frac{1}{2}b(t)\sin\theta \tag{4.55}$$

Multiplying $s_{QAM}(t)$ by $c_q(t)$ followed by a low-pass filter generates the quadrature component

$$y_q(t) = \frac{1}{2}b(t)\cos\theta + \frac{1}{2}a(t)\sin\theta \tag{4.56}$$

Equations (4.55) and (4.56) show the effect of phase error in the demodulation of QAM signals. Not only is the amplitude of the desired signal component reduced by the factor $\cos\theta$, but also there is crosstalk interference to the in-phase component from the quadrature component and vice-versa.

To effect a locally generated carrier synchronized with the incoming carrier, the incoming carrier must be extracted from the received signal. It is not intuitive, however, how such extraction can be achieved. In DSBSC systems modulated with random equiprobable data, the resulting transmitted spectrum is continuous, containing no discrete carrier component. Further, the received spectrum is contaminated with noise and possibly unwanted interference. The process of extracting a carrier from such a spectrum is referred to as *carrier recovery,* and many ingenious ways have been devised to do this. In this section, we shall briefly review some of these methods.

There are two conceptually different approaches to the problem of carrier recovery. One is to simply change the rules of the game by adding the carrier as a discrete component to the modulated signal and in the demodulator filtering it out and locking onto it with a phase lock loop. This approach, though straightforward, is at the expense of the energy per bit transmitted and is thus not often employed. The other approach is to extract the carrier from the continuous received spectrum despite its obscurity. And as we recall from above, we need not only to recreate the received carrier frequency, but also we need the exact phase. Traditionally, the receiver local oscillator(s) is/are adjusted adaptively to match the frequency and phase of the received signal. There are many methods for achieving carrier frequency and phase recovery and a number of these are well covered in [9]. In modern systems, these implementations are carried out mainly in the digital domain, and this approach is well addressed in [10]. Following we will take a high-level view of one approach, the so-called *decision directed* method, which has found high favor in QAM systems, bearing in mind that wireless transport systems employ QAM predominantly.

The multiply-filter-divide method [5] and the Costas loop method [5] are commonly used carrier frequency and phase recovery methods used with BPSK and QPSK where small phase errors in the recovered carrier do not seriously degrade performance. However, for high-order QAM where, due to the compactness of the signal states, even small phase errors can impair performance, the decision directed method is much more suitable and thus preferred. As the analysis of the operation and performance of this method (and variations thereof) is highly complex, it will

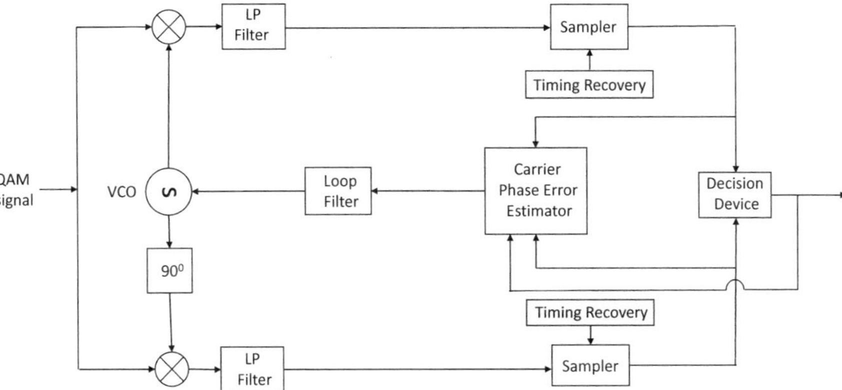

Fig. 4.28 Block diagram of a QAM receiver with decision directed carrier phase correction

not be presented here. A broad overview of the concepts behind the method, however, follows.

If frequency offset is very low, then it is possible to achieve symbol synchronization without knowledge of the carrier phase. In the decision directed method, the general approach is to assume that a recovered carrier with low frequency offset and close in phase to the desired is available, and thus, synchronized symbols are available. By low offset, we mean one less than the symbol rate, because comparisons are performed on symbols at the symbol rate. With this assumption, it makes decisions on the I and Q data symbols as if the phase error is zero. The outputs of the decision threshold units are fed to a comparison circuit, and the phase difference between the decision device output symbols and the sampler output symbols is used to generate an error signal that adjusts the phase of the local oscillator in such a way as to reduce the error, leading to phase error elimination. Figure 4.28 shows a highly simplified block diagram of a QAM receiver with decision directed carrier phase correction.

To aid in visualizing decision directed functioning, consider in Fig. 4.29 the decoded states of a 16-QAM system created by the sampler outputs shown in circles and the ideal states generated via the decision threshold units shown as black points. Noise causes the decoded constellation states created by the sampler outputs to be not precise points, but to be spread out in a disc shaped region around an average position. The decoded states are rotated from ideal by ε radians. The situation shown indicates the need for a clockwise rotation of the decoded signal constellation. The constellation comparator therefore generates an error signal that is low-pass filtered and applied to the carrier recovery *voltage-controlled oscillator* (VCO) that adjusts the carrier frequency and phase slowly so as to rotate the decoded constellation in a clockwise direction until ε approaches zero. It is rotated slowly because this improves performance in the presence of noise. In situations where the signal-to-noise ratio is low, the overall symbol error rate will obviously be larger and hence the determined phase error less reliable. One approach to minimize this negative effect is

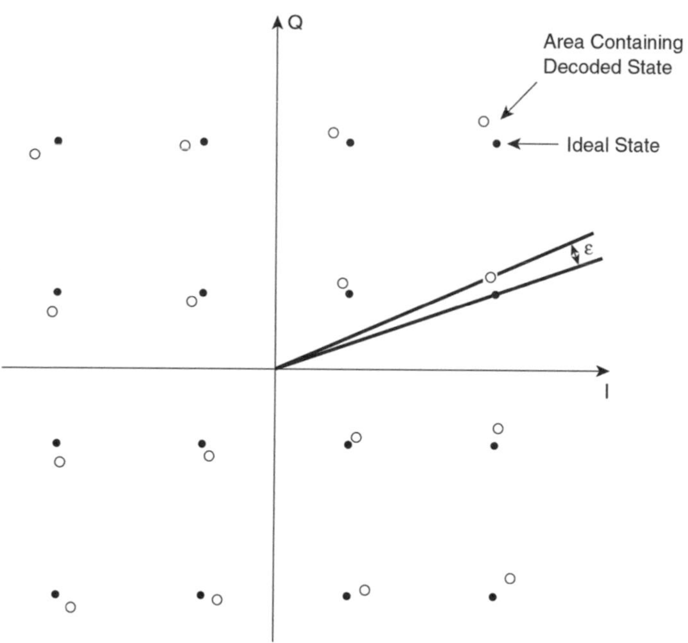

Fig. 4.29 The impact of an ε° phase error on a 16-QAM constellation

to use only the four corner symbols in rectangular constellations or giving them more weight than the others as corner symbols have a lower probability of error than the rest as they are adjacent to less symbols. For cross-constellations, the eight furthest corner symbols would be used or given more weight than the others.

Modern oscillators used for coherent modulation and demodulation and those used for up- and downconversion if applied are relatively stable and have thus phases that change relatively slowly to each other. Nonetheless, some initial carrier synchronization may be required before decision directed correction can commence. To assist the acquisition process and to increase the capture range of the loop, frequency sweeping is often used. Here, the VCO is initially fed a sweeping voltage until symbol recovery commences; then, control of the VCO is taken over by the decision directed error voltage. The decision directed method results in a fourfold ambiguity in recovered carrier phase for $(L \times L)$-QAM systems. This is because, due to the fourfold symmetry of their constellations, the constellation comparator has no way of knowing which of the four possible phases that result in a locked mode is correct. This shortcoming can, however, be overcome by using differential encoding/decoding [5]. Differential encoding operates by representing some or all of the information bits that define a signaling state as a change in the phase of the transmitted carrier, not as a component in defining an absolute location of the carrier in the constellation diagram. In the demodulator, the operation of the differential decoder is simply the inverse of the encoder. Thus, absolute phase information is unnecessary, and any phase ambiguity introduced via carrier recovery is of no consequence.

4.5.3   Timing Recovery

The recovered I and Q signals in a receiver following coherent detection must be sampled at the symbol rate and correct phase so that samples are taken at the center of the symbol period. Thus, the samplers in the demodulator discussed above require a clock synchronized to the start and stop instances of the incoming symbols in order to trigger the sample instants. Recovering this clock in the receiver is referred to as *timing recovery*. As with carrier recovery, there are many methods for achieving timing recovery, and a number of these are well covered in [9]. In modern systems, these implementations are carried out mainly in the digital domain, and this approach is well addressed in [10]. Following, we will take a high-level view of the so-called *square and filter method*. Such a method is fast, independent of carrier frequency offset and of modulation symbol level, and robust under fading channel conditions and relies only on the quasi-stationary nature of a signal after it is squared. As a result, it and variations thereof have found high favor in QAM systems, bearing in mind that wireless transport systems employ QAM predominantly.

A block diagram of a timing recovery circuit using the square and filter method is shown in Fig. 4.30a. Here, the clock is extracted from the detected I or Q data streams at the output of the low pass filter following coherent detection. Since the spectra of such signals are continuous, typically filtered $\sin x/x$ in shape and containing no discrete $1/\tau_B$ symbol rate component, a nonlinear operation on it is required to extract the symbol rate frequency. The input signal is fed to a squaring

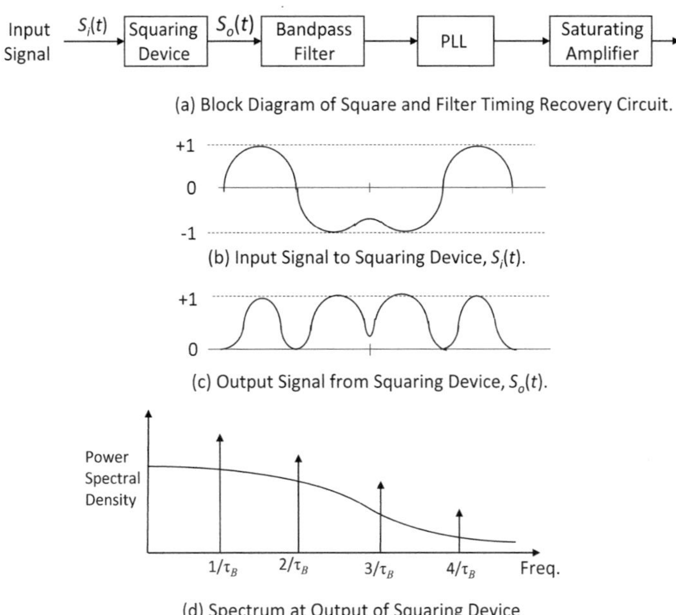

(a) Block Diagram of Square and Filter Timing Recovery Circuit.

(b) Input Signal to Squaring Device, $S_i(t)$.

(c) Output Signal from Squaring Device, $S_o(t)$.

(d) Spectrum at Output of Squaring Device

Fig. 4.30 Square and filter timing recovery method

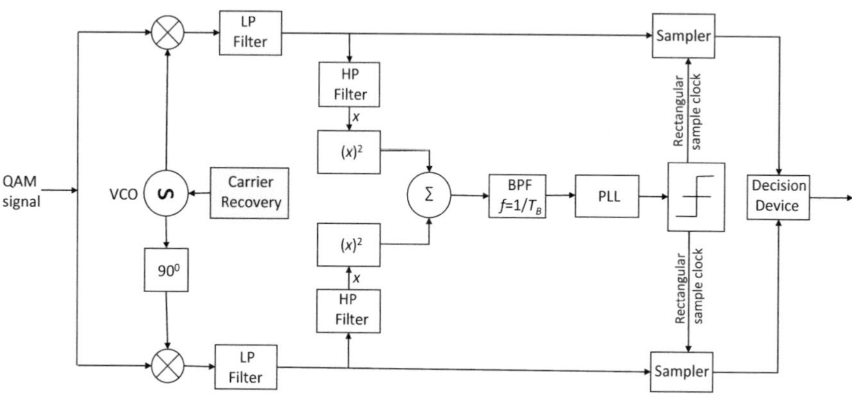

Fig. 4.31 Symbol timing recovery method is a QAM system

device that performs the nonlinear operation. Shown in Fig. 4.30b is an input signal $S_i(t)$ to a squaring device and in Fig. 4.30c the output $S_o(t)$ of that device. We observe that $S_o(t)$ has repetitive positive amplitude peaks that always correspond, with a time delay, with the input symbol transitions. It is not surprising, therefore, that $S_o(t)$ contains a Fourier component at the fundamental frequency of the data clock. In fact, it can be shown mathematically that the result of squaring the input signal is to create a spectrum as shown in Fig. 4.30d, which, though it continues to have a continuous component, also contains discrete spectral lines at multiples of $1/\tau_B$. A bandpass filter following the squarer is centered at $1/\tau_B$, and its output is fed to a phase-locked loop (PLL) that locks onto it and outputs a sinusoidal signal. This signal is typically fed to a high gain saturating amplifier which outputs a square wave signal similar to the original data clock. Despite the filtering effects of the bandpass filter and the PLL, it is not possible to totally eliminate the noise and continuous spectrum immediately surrounding the desired spectral component. As a result, the recovered timing signal contains an error component. This is referred to as timing jitter and, if significant, can degrade system performance.

In Fig. 4.31, we show a version of the square and filter method as applied to a QAM receiver. Here, information from both the I and Q rails is utilized in the clock recovery process. The high-pass filters shown is necessary to remove data pattern dependent jitter. By summing the signals out of the squaring devices the signal-to-noise ratio of the signal into the bandpass filter and the PLL is improved and hence timing jitter is reduced. In modern systems, such recovery is usually digitized.

4.6   Summary

The modulation methods used in most 5G wireless backhaul links are mainly various versions of quadrature amplitude modulation (QAM). At the low end of modulation order, we find BPSK (not QAM) and QPSK, which is really 4-QAM. At the high end, we find 8192-QAM and in a few instances 16,348-K, often referred to as 16 K-QAM. These methods, therefore, are the ones we studied in this chapter, along with a number of schemes involved in their realization such as up- and downconversion and carrier and symbol timing recovery.

Acknowledgement The author wishes to thank Sung-Moon Michael Yang for his valuable help and advice during the preparation of the carrier and timing recovery sections of this chapter.

References

1. Feher K (1981) Digital communications: microwave applications. Prentice-Hall, Upper Saddle River
2. Lucky RW, Salz J, Weldon EJ (1968) Principles of data communication. McGraw-Hill, New York
3. Bennett WR, Davey JR (1965) Data transmission. McGraw-Hill, New York
4. Taub H, Schilling DL (1971) Principles of communication systems. McGraw-Hill, New York
5. Morais DH (2004) Fixed broadband wireless communications: principles and practical applications. Prentice-Hall PTR, Upper Saddle River
6. Burr A (2001) Modulation and coding for wireless communications. Prentice-Hall, Harlow
7. Morais DH, Inventor (2013) Quadrature amplitude modulation via modified-square signal point constellation. United States Patent, Patent No. US 8,422,579 B1, April 16, 2013
8. Morais DH, Inventor (2014) Hard and soft bit demapping for QAM non-square constellations. United States Patent, Patent No. US 8,718,205 B1, May 6, 2014
9. Proakis J, Salehi M (2008) Digital communications, 5th edn. McGraw-Hill, New York
10. Yang SM (2020) Modern digital radio communication signals and systems, 2nd edn. Springer, Cham

Chapter 5
Performance Optimization Techniques

5.1 Introduction

The probability of error equations reviewed in Chap. 4 assumed an ideal linear path between the transmitter and receiver. However, as was seen in Chap. 3, real-world terrestrial transmission often deviates from this ideal. Digital wireless systems employing linear modulation methods are particularly susceptible to the in-band distortions created via multipath fading. Further, this susceptibility increases, in general, as the number of modulation states increases. Because spectrum is limited, there is constant pressure to improve spectral efficiency as a way to increase throughput. This in turn leads to systems with higher and higher numbers of modulation states. Such systems, in addition to being highly susceptible to in-band distortion, are also susceptible to their own implementation imperfections. This makes the attainment of error rate performance close to ideal difficult to achieve, even in a linear transmission environment, and results in a bit error rate (BER), at even high signal-to-noise ratios, that levels off at a residual value that may be higher than desirable. A number of highly effective techniques have been developed to address these susceptibilities. As a result, by their application, the transmission of very high data rates at very high levels of spectral efficiency is possible. In this chapter, some of the more important of these techniques that are or may be applied in wireless transport links are reviewed, including forward error correction (FEC), adaptive modulation and coding (AMC), power amplifier linearization, phase noise suppression, quadrature modulation/demodulation imperfection mitigation, and adaptive equalization.

5.2   Forward Error Correction Coding

5.2.1   Introduction

Coding, in the binary communications world, is the process of adding a bit or bits to useful data bits in such a fashion as to facilitate the detection or correction of errors incurred by such useful bits as a result of their transmission over a non-ideal channel. Such a channel, for example, may be one that adds noise, or interference, or unwanted nonlinearities. In this section, we focus on error correction codes, including the very important *low-density parity-check* (LDPC) codes, as well as the tried-and-true *Reed-Solomon* codes as these often find application in wireless transport systems. In addition, we take a look at the recently introduced *polar* codes.

Error-control coding is a means of permitting the robust transmission of data by the deliberate introduction of redundancies into the data creating a codeword. One method of accomplishing this is to have a system that looks for errors at the receiving end, and once an error is detected, makes a request to the transmitter for a repeat transmission of the codeword. In this method, called *Automatic Repeat Request* (ARQ), a return path is necessary. Error correction coding that is not reliant on a return path inherently adds less delay to transmission. Such coding is referred to as *forward error correction* (FEC) coding. For digitally modulated signals, detected in the presence of noise, use of FEC results in the reduction of the residual BER, usually by several orders of magnitude, and a reduction of the receiver 10^{-6} threshold level by about one to several dBs depending on the specific scheme employed. Figure 5.1 shows typical error performance characteristics of an uncoded

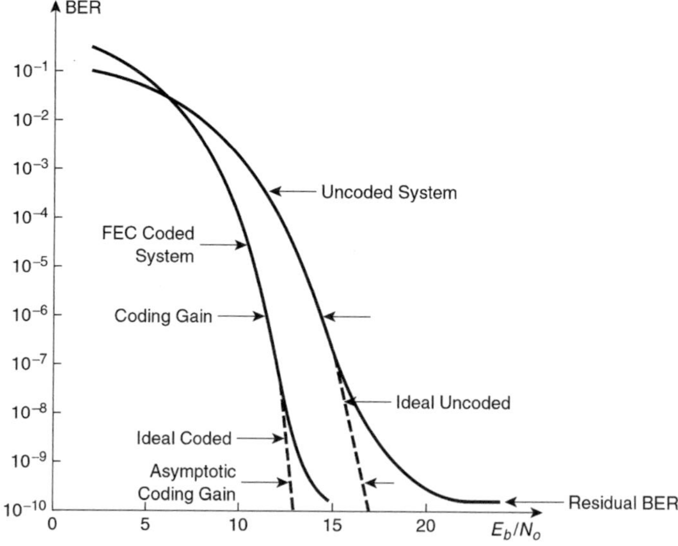

Fig. 5.1 Typical error performance of an uncoded versus FEC coded system

versus FEC coded digitally modulated system. The advantage provided by a coded system can be quantified by *coding gain*. The coding gain provided by a particular scheme is defined as the reduction in E_b/N_0 in the coded system compared to the same system but uncoded for a given BER and the same data rate. Coding gain varies significantly with BER, as can be seen from Fig. 5.1, and above a very high level may even be negative. As BER decreases, the coding gain increases until it approaches a limit as the BER approaches zero (zero errors). This upper limit is referred to as the *asymptotic coding gain* of the coding scheme.

FEC works by adding extra bits to the bitstream prior to modulation according to specific algorithms. These extra bits contribute no new message information. However, they allow the decoder, which follows the receiver demodulator, to detect and correct, to a finite extent, errors as a result of the transmission process. Thus, improvement in BER performance is at the expense of an increase in transmission bit rate. The simplest error detection-only method used with digital binary messages is the parity-check scheme. In the even-parity version of this scheme, the message to be transmitted is bundled into blocks of equal bits and an extra bit is added to each block so that the total number of 1 s in each block is even. Thus, whenever the number of 1 s in a received block is odd the receiver knows that a transmission error has occurred. Note, however, that this scheme can detect only an odd number of errors in each block. For error detection and correction, the addition of several redundant (check) bits is required. The number of redundant bits is a function of the number of bits in error that are required to be corrected.

FEC codes can be classified into two main categories, namely, *convolution codes*, where the message bitstream is encoded in a continuous fashion, and *block codes*, where the message bitstream is split into fixed-length blocks and the encoder adds check bits to each block. Convolution codes are powerful ones and have been much employed. However, recently they are being supplanted by two classes of block codes that provide operation closer to the ideal, namely, low-density parity-check (LDPC) codes and the recently introduced polar codes. Another class, Reed-Solomon codes, has been around for a long time and its use persists due to its ease of implementation, relative effectiveness in the presence of burst errors, and relatively low latency. Here, we will study LDPC and Reed-Solomon codes as these are currently the most often used in point-to-point wireless transport links in support of mobile networks. We will also, in addition, look at polar codes as, given their performance advantages, it is not unreasonable to assume that they may find application in wireless transport links in the future. We will first look at block coding, in general, and then turn our attention to LDPC, Reed-Solomon, and finally polar codes.

5.2.2   Block Codes

In *systematic binary linear* block encoding, the input bitstream to be encoded is segregated into sequential message blocks, each k bits in length. The encoder adds

Table 5.1 A (5,2) Block Code

| | Codeword | |
Codeword #	Message bits	Check bits
1	00	000
2	01	110
3	10	011
4	11	101

r check bits to each message block, creating a codeword of length n bits, where $n = k + r$. The codeword created is called an (n, k) block codeword, having a block length of n and a coding rate of k/n. Such a code can transmit 2^k distinct codewords, where, for each codeword, there is a specific mapping between the k message bits and the r check bits. The code is *systematic* because, for all codewords, a part of the sequence in the codeword (usually the first part) coincides with the k message bits. As a result, it is possible to make a clear distinction in the codeword between the message bits and the check bits. The code is *binary* because its codewords are constructed from bits, and *linear* because each codeword can be created by linear *modulo-2 addition* of two or more other codewords. Modulo-2 addition is defined as follows:

$$0 + 0 = 0$$
$$0 + 1 = 1$$
$$1 + 0 = 1$$
$$1 + 1 = 0$$

The following simple example will help explain the basic principles involved in linear binary block codes.

Example 5.1: The Basic Features and Functioning of a Simple Linear Binary Code

Consider a (5, 2) block code where 3 check bits are added to a 2-bit message. There are thus four possible messages and hence four possible 5-bit encoded codewords. Table 5.1 shows the specific choice of check bits associated with the message bits. A quick check will confirm that this code is linear. For example, codeword 1 can be created by the modulo-2 addition of codewords 2, 3, and 4. How does the decoder work? Suppose codeword 3 (10011) is transmitted, but an error occurs in the second bit so that the word 11011 is received. The decoder will recognize that the received word is not one of the four permitted codewords and thus contains an error. This being so, it compares this word with each of the permitted codewords in turn. It differs in four places from codeword 1, three places from codeword 2, one place from codeword 3, and two places from codeword 4. The decoder therefore concludes that it is codeword 3 that was transmitted, as the word received differs from it by the least number of bits. Thus, the decoder can detect and correct an error.

The number of places in which two words differ is referred to as the *Hamming distance*. Thus, the logic of the decoder in Example 5.1 is, for each received word, select the codeword closest to it in Hamming distance. The minimum Hamming distance between any pair of codewords, d_{min}, is referred to as the *minimum distance of the code*. It provides a measure of the code's minimum error-correcting capability and thus is an indication of the code's strength. In general, the error-correcting capability, t, of a code, is defined as the maximum number of guaranteed correctable errors per codeword, is given by

$$t = \left\lfloor \frac{d_{min} - 1}{2} \right\rfloor \tag{5.1}$$

where $\lfloor i \rfloor$ means the largest integer not to exceed i.

An important subcategory of block codes is *cyclic block codes*. A code is defined as cyclic if any cyclic shift of any codeword is also a codeword. Thus, for example, if 101101 is a codeword, then 110110 is also a codeword, since it results from shifting the last-bit to the first-bit position and all other bits to the right by one position. This subcategory of codes lends itself to simple encoding algorithms. Further, because of their inherent algebraic format, decoding is also accomplished with a simple structure.

Block decoding can be accomplished with *hard decision decoding*, where the demodulator outputs either ones or zeros as in Example 5.2. Here, the codeword chosen is the one with the least Hamming distance from the received sequence. However, decoding can be improved by employing *soft decision decoding*. With such decoding, the demodulator output is normally still digitized, but to greater than two levels, typically eight or more. Thus, the output is still "hard", but more closely related to the analog version and thus contains more information about the original sequence. Such decoding can be accomplished in a number of ways. One such way is to choose as the transmitted codeword the one with the least *Euclidian distance* from the received sequence. The Euclidian distance between sequences is, in effect, the root mean square error between them. To demonstrate the advantage of soft decision-based decoding over Hamming distance-based decoding, a simple example of decoding via both methods is presented below.

Example 5.2: Demonstration of the Advantage of Soft Decision Decoding via Euclidian Distance over Hard Decision Decoding via Hamming Distance

Consider an encoder that produces the four codewords in Table 5.1. This is a block encoder and it serves the purpose of conveying in a straightforward fashion the basic concept and advantage of decoding using Euclidean distance versus Hamming distance as the decoding metric. Assume that codeword 2 (01110) is sent over a noisy channel in the form of the signal shown in Fig. 5.2a, and, as a result, the demodulator analog output signal is as shown in

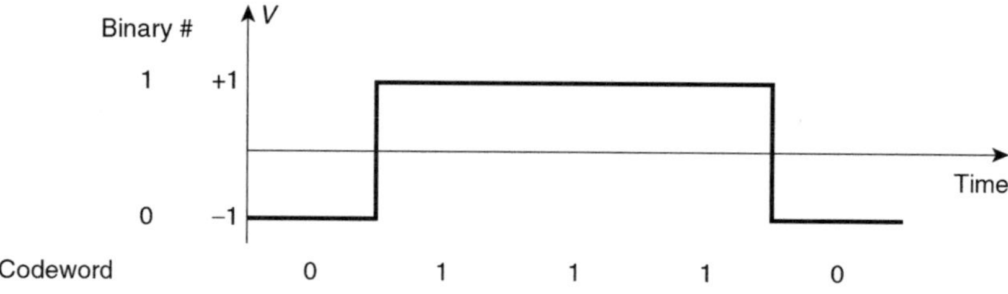

(a) Signal Sent Over Noisy Channel Representing Codeword 01110

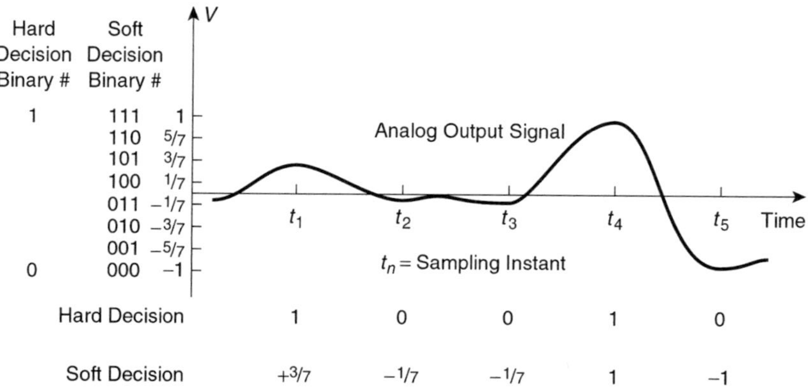

(b) Demodulated Analog Output Signal Resulting from Reception of Signal in (a)
Plus Noise and Resulting Hard and Soft Decisions

Fig. 5.2 Examples of hard and soft decoding

Fig. 5.2b. This analog output leads to a hard decision output of 10010 and, with eight-level quantization, a soft decision output as indicated on the figure.

Let's first assume that decoding is based on hard decisions. If $d_H(r, n)$ represents the Hamming distance between the hard decision outputs of the received signal and codeword n, then simple comparison yields $d_H(r,1) = 2$, $d_H(r,2) = 3$, $d_H(r,3) = 1$, and $d_H(r,4) = 4$. Since $d_H(r,3)$ is the smallest Hamming distance, the decoder declares that the received codeword is codeword 3, i.e., 10011. It thus decodes in error.

Let's now assume that the decoder is using soft decisions and that $d_E(r,n)$ represents the Euclidean distance between the soft decision outputs of the received signal and codeword n. We compute the squared Euclidean distance, $d_E^{\,2}(r,1)$, by determining the error between the soft decision output and codeword 1 for each of the five bits sent, squaring these errors, and then adding the squared values together. Since codeword 1 is 00000, its true output per bit would be $-1, -1, -1, -1, -1$, and thus the errors between its potential bit outputs and the received signal soft outputs are, sequentially, 1 3/7, 6/7, 6/7, 2, and 0. Thus, $d_E^2(r,1)$ is given by

$$d_E^2(r, 1) = (1\ 3/7)^2 + (6/7)^2 + (6/7)^2 + (2)^2 + (0)^2 = 7.51$$

Applying this same process to the other three codewords, we get

$$d_E^2(r, 2) = (1\ 3/7)^2 + (1\ 1/7)^2 + (1\ 1/7)^2 + (0)^2 + (0)^2 = 4.65$$
$$d_E^2(r, 3) = (4/7)^2 + (6/7)^2 + (6/7)^2 + (0)^2 + (2)^2 = 5.80$$
$$d_E^2(r, 4) = (4/7)^2 + (1\ 1/7)^2 + (1\ 1/7)^2 + (2)^2 + (2)^2 = 10.94$$

Since $d_E^2(r,2)$ is the smallest squared Euclidean distance, then $d_E(r,2)$ is the smallest Euclidean distance and hence the encoder chooses codeword 2, thus making the correct decision. In effect, the decoder's decision is based on the fact that it can't, with much confidence, decide what are the first three bits that have been sent, but it can with high confidence decide that the last two bits sent are 10. Since only codeword 2 had these last two bits, it decides, correctly, that codeword 2 was sent.

Another and much-used form of soft decision decoding is via *logarithmic likelihood ratios* (LLRs) where each received bit is processed as the probability, in a logarithmic form, of it being either a 0 or a 1.

5.2.3 Classical Parity-Check Block Codes

Before we describe the features of LDPC codes, we review some of the features of classical *parity-check block codes*. In such a code, each codeword is of a given length, n say, contains a given number of information bits, k say, and a given number of *parity-check bits*, r say, and thus $r = n - k$. The structure can be represented by a *parity-check matrix* (PCM), where there are n columns representing the digits in the codeword, and r rows representing the equations that define the code. Consider one such code, where the length n is 6, the number of information bits $k = 3$, and hence, the number of parity bits r is 3. The rate of this code is thus $k/n = 3/6 = 1/2$. We label the information bits V1 to V3 and the parity bits V4 to V6. The parity-check equations for this code are shown in Eq. (5.2) below, where $+$ represents modulo-2 addition:

$$
\begin{aligned}
V1 + V2 + V4 \qquad\quad &= 0 \\
V2 + V3 + V5 \qquad\quad &= 0 \\
V1 + V2 + V3 + V6 &= 0
\end{aligned}
\tag{5.2}
$$

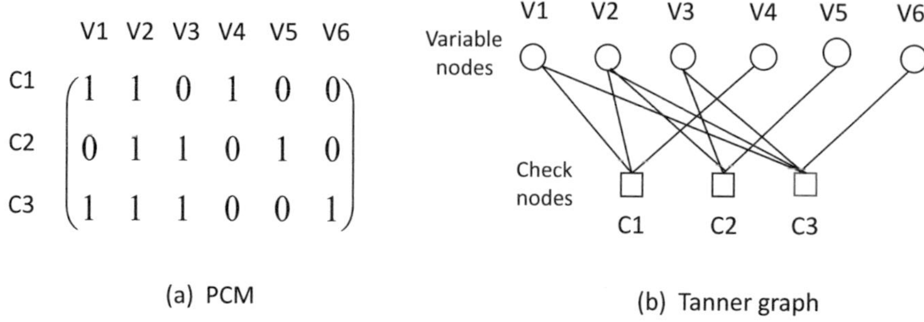

Fig. 5.3 A PCM and associated Tanner graph. (From [1] with the permission of Springer)

The above equations can be represented in matrix form, as shown below in Fig. 5.3a [1], where each equation maps to a row of the matrix. This matrix is referred to as the PCM associated with Eq. (5.2).

Equation 5.2 can also be represented in graphical form. When done, such a graph is referred to as a *Tanner graph*. A Tanner graph is a bipartite graph, i.e., a graph which contains nodes of two different types, and lines (also referred to as edges) which connect nodes of different types. The bits in the codeword form one set of nodes, referred to as *variable nodes* (VNs), and the parity-check equations that the bits must satisfy form the other set of nodes, referred to as the *check nodes* (CNs). The Tanner graph corresponding to the PCM matrix above is shown in Fig. 5.3b.

Errors can be detected within limits in any received codeword by simply checking if it satisfies all associated parity-check equations. However, block codes can only detect a set of errors if errors don't change one codeword into another. Further, even if this is not the case, they can only detect bit errors if the number of these errors is less than the minimum distance, d_{min}.

To not only detect bit errors but also correct them, the decoder must determine which codeword was most likely sent. One way to do this is to choose the codeword closest in minimum distance to the received codeword. This method of decoding is called *maximum likelihood* (ML) *decoding*. For codes with a short number of information bits this approach is feasible as the computation required is somewhat limited. However, for codes with thousands of information bits in a codeword, the computation required becomes too excessive and expensive. For such codes, alternative decoding methods have been devised and will be discussed below.

5.2.4   *Low-Density Parity-Check (LDPC) Codes*

Low-density parity-check (LDPC) codes are linear FEC codes and were first proposed by Gallager [2] in his 1962 Ph. D. thesis. They can provide higher coding gains and lower error floors than convolution turbo codes, and in the decoding process be computationally more efficient. LDPC codes are distinguished from

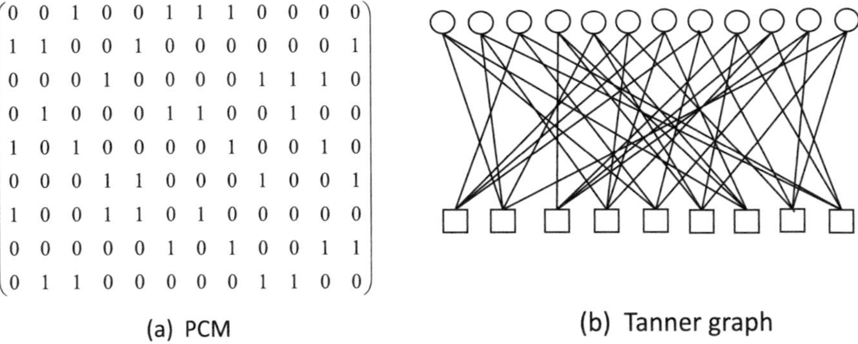

(a) PCM   (b) Tanner graph

Fig. 5.4 PCM and Tanner graph for an LDPC code where $n = 12$. (From [1], with the permission of Springer)

other parity-check codes by having parity-check matrices where the percentage of 1 s is low, i.e., of low density, hence the nomenclature. This sparseness of 1 s results in not only a decoding complexity which increases only linearly with code length but also a minimum distance which increases only linearly with code length.

LDPC codes are said to be either regular or irregular. An LDPC code is regular if all the VNs have the same degree, i.e., they are connected to the same number of CNs, and all the CNs have the same degree, i.e., they are connected to the same number of VNs. When this is the case, then every code bit is contained in the same number of equations, and each equation contains the same number of code bits. Irregular codes relax these conditions, allowing VNs and CNs of different degrees. Irregular codes, it has been found, can provide better performance than regular ones.

Figure 5.4a [1] shows the PCM of a simple LDPC code where $n = 12$, and Fig. 5.4b shows the associated Tanner graph. It will be noted that this is a regular LDPC code, wherefrom the PCM perspective each code bit is contained in 3 equations and each equation involves 4 code bits, and from the Tanner graph perspective, each bit node has 3 lines connecting it to parity nodes and each parity node has 4 lines connecting it to bit nodes. We note that in the PCM there are 108 positions in all of which only 36, or 33%, are ones.

5.2.4.1   Encoding of Quasi-Cyclic LDPC Codes

There exist a class of LDPC codes called *Quasi-cyclic* (QC) LDPC codes [3]. With these codes, encoding and decoding hardware implementation tends to be easier than with other types of LDPC codes, achieving this without measurably degrading the relative performance of the code. The PCM of a QC LDPC code is defined by a small graph, called a *base graph* or *protograph*. The base graph, U say, is transformed into the PCM, H say, by replacing each entry in U with a cyclically shifted to the right version of a Z x Z identity matrix, I say. Here, Z is referred to as the *lifting factor*, as the larger it is, the larger the size of H. Base graph entries are not binary, but rather range from -1 to $(Z - 1)$. By convention, -1 means a matrix with all 0 entries. The entries 0 to $(Z - 1)$ represent the possible cyclically shifted versions of I. Example 5.4 below will demonstrate the construction of a QC LDCP code.

Example 5.3: Construction of a QC LDPC Code

Consider a QC LDPC code where the Identity matrix I is as below:

$$I = \begin{pmatrix} 1 & 0 & 0 \\ 0 & 1 & 0 \\ 0 & 0 & 1 \end{pmatrix}$$

Then the base graph entries are given by:

$$-1 = \begin{pmatrix} 0 & 0 & 0 \\ 0 & 0 & 0 \\ 0 & 0 & 0 \end{pmatrix} \quad 0 = \begin{pmatrix} 1 & 0 & 0 \\ 0 & 1 & 0 \\ 0 & 0 & 1 \end{pmatrix} \quad 1 = \begin{pmatrix} 0 & 1 & 0 \\ 0 & 0 & 1 \\ 1 & 0 & 0 \end{pmatrix} \quad 2$$

$$= \begin{pmatrix} 0 & 0 & 1 \\ 1 & 0 & 0 \\ 0 & 1 & 0 \end{pmatrix} \quad 3 = \begin{pmatrix} 1 & 0 & 0 \\ 0 & 1 & 0 \\ 0 & 0 & 1 \end{pmatrix}$$

Thus, if we have a base graph U given by:

$$U = \begin{pmatrix} 3 & -1 & -1 & 1 \\ -1 & 0 & -1 & 2 \end{pmatrix}, \quad \text{then} \, H$$

$$= \begin{pmatrix} 1 & 0 & 0 & 0 & 0 & 0 & 0 & 0 & 0 & 0 & 1 & 0 \\ 0 & 1 & 0 & 0 & 0 & 0 & 0 & 0 & 0 & 0 & 0 & 1 \\ 0 & 0 & 1 & 0 & 0 & 0 & 0 & 0 & 0 & 1 & 0 & 0 \\ 0 & 0 & 0 & 1 & 0 & 0 & 0 & 0 & 0 & 0 & 0 & 1 \\ 0 & 0 & 0 & 0 & 1 & 0 & 0 & 0 & 0 & 1 & 0 & 0 \\ 0 & 0 & 0 & 0 & 0 & 1 & 0 & 0 & 0 & 0 & 1 & 0 \end{pmatrix}$$

We note that the key to the high performance of a QC LDPC code is the construction of the base graph. Though the number of identity matrix cyclic permutations is Z, in practice the number of permutations used is restricted to simplify implementation.

5.2.4.2   Decoding of LDPC Codes

A big distinguishing feature between LDPC codes and classical block codes is how they are decoded. Unlike classical codes that are usually of short block length and decoded via ML decoding, LDPC codes are usually of long block length and decoded iteratively.

LDPC codes are decoded using *message-passing* algorithms [4] since their functioning can be described as the passing of messages along the lines of the Tanner graph. Each node on the Tanner graph works in isolation, having access only to the information conveyed by the lines connected to it. The message passing algorithms create a process where the messages pass back and forth between the bit nodes and check nodes iteratively. For optimum decoding the messages passed are estimates of the probability that the codeword bit information passed is 1. Each estimate is in the form of a *logarithmic likelihood ratio* (LLR), where, for a codeword bit b_i, $LLR(b_i) = \ln \text{prob.}(b_i = 0) - \ln \text{prob.}(b_i = 1)$. A positive LLR indicates a greater confidence that the associated bit is of value 0, while a negative LLR indicates a greater confidence that the bit value is 1. The magnitude of the LLR expresses the degree of confidence. Decoding as described above is termed *belief propagation* decoding and proceeds as follows:

1. Each codeword is outputted from the channel not as hard outputs (1 s or 0 s), but rather as soft outputs. These soft outputs are converted into initial estimates in the form of LLRs.
2. Each bit node sends its initial estimate to the check nodes on the lines connected to it.
3. Each check node makes new estimates of the bits involved in the parity equation associated with that node and sends these new estimates via the connecting lines back to the associated bit nodes.
4. New estimates at the bit nodes are sent to the check nodes and process steps 3 and 4 repeated until a permitted codeword is found or the maximum number of permitted iterations reached.

Many decoding iterations may be necessary before there is a convergence of the estimates and all the errors are corrected. The larger the block length, the larger the confidence level in the error-correcting capability; hence, large block lengths are desirable. However, large block lengths increase computational complexity as this increases the size of the parity-check matrix and thus increases the number of decoding calculations required to estimate each bit. Thus, the result of large block lengths coupled with several decoding iterations is increased latency.

5.2.5 Reed-Solomon (RS) Codes

Reed-Solomon codes [5] were developed in 1960 by I. Reed and G. Solomon and have been widely used in digital wireless systems. These codes are "nonbinary" cyclic block codes. In nonbinary block codes, the input bitstream is converted into symbols m bits long and these symbols are segregated into message blocks, each k symbols in length. The encoder then adds r check symbols, each also m bits long, creating a codeword of length n symbols. RS (n, k) codes exist for all n and k for which

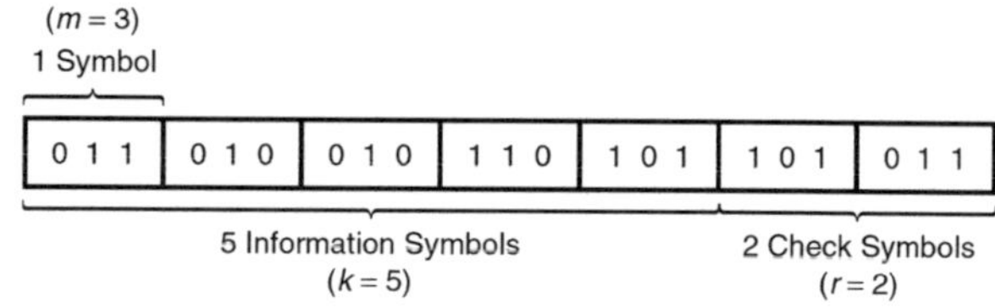

$$0 < k < 2^m + 2 \tag{5.3}$$

However, for the most commonly used RS (n,k) code,

$$(n, k) = (2^m - 1, 2^m - 1 - 2t) \tag{5.4}$$

where t is the number of correctable symbol errors per codeword, and thus the number of parity symbols, $n - k$, equals 2 t.

Figure 5.5 shows an example of an RS codeword based on Eq. (5.4) and where symbol length is 3 bits ($m = 3$) and $t = 1$ and hence $n = 7$ and $k = 5$.

For nonbinary codes, the distance between two codewords is defined as the number of places in which the symbols differ. For RS codes, the code minimum distance $d_{\min}$ is given by [6] to be

$$d_{\min} = n - k + 1 \tag{5.5}$$

Substituting $d_{\min}$ from Eq. (5.5) into Eq. (5.1), we get

$$t = \left\lfloor \frac{n - k}{2} \right\rfloor \tag{5.6a}$$

$$= \left\lfloor \frac{r}{2} \right\rfloor \tag{5.6b}$$

where $\lfloor i \rfloor$ means the largest integer not to exceed i.

Reed-Solomon codes may be shortened by making a number of information designated symbols at the encoder zero, the first i symbols say, not transmitting them, and then reinserting them at the decoder prior to decoding. A $(n - i, k - i)$ *shortened code* of the original code is created in such a fashion, with the same minimum distance, and hence the same correctable symbol error capability, of the original code.

RS codes achieve the largest possible $d_{\min}$ and, hence, the largest possible error-correcting capability of any linear code, given the same values of n and k. Further, they are especially effective in correcting long strings of errors normally referred to as burst errors which can be caused by deep fading in a communications channel. This burst error-correcting capability is because for a given symbol being in error, the bit error correction performance of the code is independent of the number of bits in error in the symbol. Consider, for example, an RS (63, 57) code where there are

Table 5.2 (7,4) Hamming code codewords

Codeword #	Data bits				Check bits		
	D_1	D_2	D_3	D_4	C_1	C_2	C_3
1	0	0	0	0	0	0	0
2	0	0	0	1	0	1	1
3	0	0	1	0	1	0	1
4	0	0	1	1	1	1	0
5	0	1	0	0	1	1	0
6	0	1	0	1	1	0	1
7	0	1	1	0	0	1	1
8	0	1	1	1	0	0	0
9	1	0	0	0	1	1	1
10	1	0	0	1	1	0	0
11	1	0	1	0	0	1	0
12	1	0	1	1	0	0	1
13	1	1	0	0	0	0	1
14	1	1	0	1	0	1	0
15	1	1	1	0	1	0	0
16	1	1	1	1	1	1	1

6 bits in a symbol. Such a code is capable of correcting any three symbols in a block of 63. Consider also what happens if an error burst of up to 13 contiguous bits occurs. These bits in error would be contained within three symbols regardless of when the sequence commenced and would thus all be corrected. Further, error bursts of between 14 and 18 contiguous bits may, depending on when the sequence commenced, be contained within three symbols. This capability provides RS codes with a significant burst error-handling advantage over binary codes and helps explain their popularity. Note, however, that this advantage comes at a price. If the bits in error had been spread over more than three symbols in the 63-symbol block, then all the symbol errors and hence all the bit errors could not have corrected.

To gain an intuitive understanding of how RS codes are mathematically generated and decoded we can accomplish this by looking at how a *Hamming Code* works mathematically, as the principles are similar but simpler in the case of the Hamming code [7]. Hamming codes are a simple class of block codes that have a minimum distance of 3 and thus, by Eq. (5.1), are capable of correcting all single errors. We will consider a (7,4) Hamming code consisting of 4 information bits and 3 check bits. Table 5.2 shows all possible codewords where, for any given sequence of data bits, the check bits are calculated via the following equations, where $+$ implies modulo-2 addition:

$$C_1 = D_1 + D_2 + D_3 \tag{5.7a}$$

$$C_2 = D_1 + D_2 + D_4 \tag{5.7b}$$

Table 5.3 Bit decoding true/false matrix

Bit in error	Eq. (5.7a)	Eq. (5.7b)	Eq. (5.7c)
None	True	True	True
D_1	False	False	False
D_2	False	False	True
D_3	False	True	False
D_4	True	False	False
C_1	False	True	True
C_2	True	False	True
C_3	True	True	False

$$C_3 = D_1 + D_3 + D_4 \tag{5.7c}$$

With a codeword consisting of 7 bits, then there are $2^7 = 128$ possible combinations of these bits. Only 16 such combinations, however, are valid codewords. This, if a codeword is received that is not one of these 16 then is clearly in error. To calculate if an incoming codeword is in error, Eqs. (5.7a), (5.7b), and (5.7c) are performed on the codeword. If no error has occurred, then the computed values of C_1, C_2, and C_3 will match those received. If one of the seven bits is in error, however, then a certain subset of the computed values of C_1, C_2, and C_3 will not match those received, i.e., the results will be false. We can thus easily compute if an error has occurred. We wish, however, not only to detect an error but also to correct it. This is done by referring to Table 5.3, which shows, for each bit in error, which equation (s) will be false. Once the bit that is in error is located, it is corrected by inverting it.

To demonstrate how this works, suppose, for example, that the following codeword was received:

$$\begin{array}{ccccccc} D_1 & D_2 & D_3 & D_4 & C_1 & C_2 & C_3 \\ 1 & 1 & 0 & 1 & 1 & 0 & 0 \end{array}$$

Performing the modulo-2 addition equations we get:

$$C_1 = 1 + 1 + 0 = 0 \quad \text{False, since } C_1 \text{ received is 1}$$

$$C_2 = 1 + 1 + 1 = 1 \quad \text{False, since } C_2 \text{ received is 0}$$

$$C_3 = 1 + 0 + 1 = 0 \quad \text{True, since } C_3 \text{ received is 0}$$

Then, by Table 5.3, bit D_2 is in error, and inverting it we get 1 0 0 1 1 0 0 which is codeword number 10, a correct codeword.

Reed -Solomon codes work essentially the same as Hamming codes, except for the fact that with RS codes we must deal with multi-bit symbols rather than

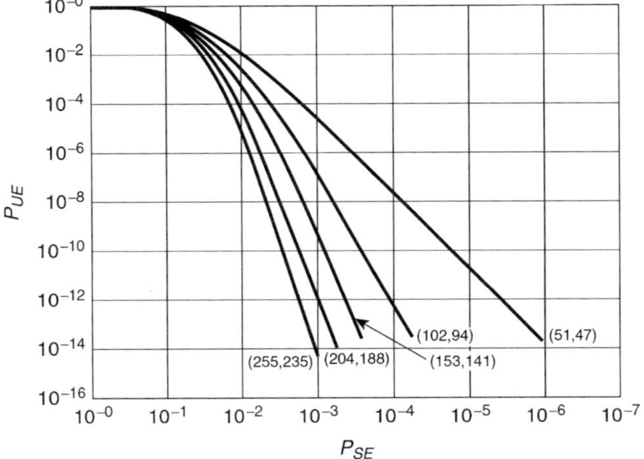

(a) Performance Curves for RS Codes of Rate 0.92

(By permission from Comtech AHA Corporation)

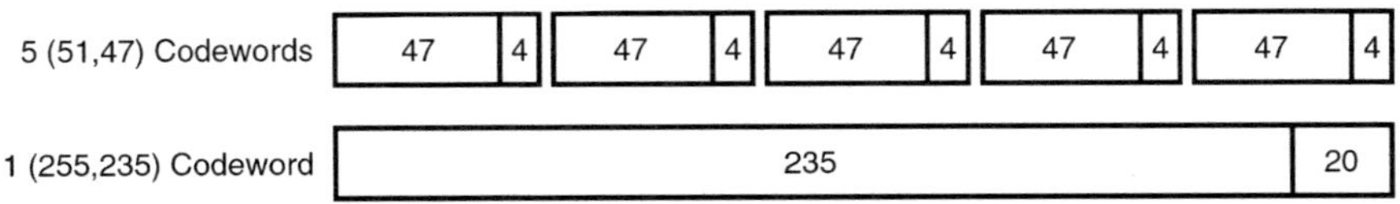

(b) Comparison of 1 (255,235) Codeword with a Sequence of 5 (51,47) Codewords

Fig. 5.6 Comparison of RS codes of rate 0.92. (a) from [7], with the permission of Advanced Hardware Architecture)

individual bits and that being more complex they require much more computational processing.

Additive Gaussian noise typically causes random errors. With random noise, the bit error probabilities are independent of each other. The longer the codeword length, the greater the probability that the number of random errors in a codeword will be the average number of errors for that length and thus the more effective the code. Thus, to combat random errors, RS codes usually have long codewords. Figure 5.6a [7] shows performance curves for RS codes of rate 0.92 and codelengths n of values varying from 51 to 255. The probability of input symbol error, P_{SE}, is shown on the horizontal axis, where P_{SE} is the probability that the channel will change a symbol during transmission of the message. The probability of an uncorrectable error, P_{UE}, is shown on the vertical axis, where P_{UE} the ratio of the number of uncorrectable code blocks to the total number of received code blocks where the number of received code blocks is very large. It will be observed that the larger the value of n, the more effective the code. The symbol length $m = 8$, and hence the codeword length $n = 255$, is a popular choice. Figure 5.6b shows this codeword as well as a sequence of five of the shortened (51, 47) codewords. Since the rates are the same, given a choice of using the (255, 235) code versus the (51, 47) option, one would

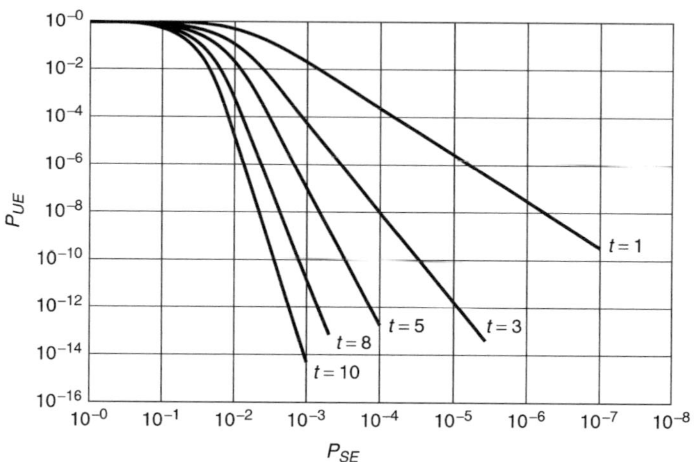

(By permission from Comtech AHA Corporation)

Fig. 5.7 Random symbol block performance for the RS(255, k) code for $k = 235$, $t = 10$, through $k = 253$, $t = 1$. (From [7], with the permission of Advanced Hardware Architecture)

clearly choose the former, assuming the added delay (latency) in the transmission is tolerable.

Figure 5.7 [7] shows random symbol error block performance for an RS (255, k) code for $t = 1, 3, 5, 8$, and 10. Note that for the $t = 10$ option, a probability of input symbol error $P_{SE} = 10^{-3}$ results in a probability of output uncorrectable error, P_{UE}, of approximately 10^{-14}. This significant probability of error reduction comes at the expense of a data rate that's only a factor of 255/235, or 8.5%, larger than the uncoded rate. In fixed wireless systems, the FEC codes used typically add no more than about 10% to the information rate as adding more, though helpful in error correction, comes at a measurable reduction to spectral efficiency.

For block codes, an appropriate way to measure bit error rate is what is referred to as the corrected bit error rate (CBER). This rate is the reciprocal of the average number of correct bits between an error bit and is given by:

$$CBER = \frac{P_{UE}}{m \times n} \tag{5.8}$$

Thus, for the case where the number of bits per symbol, m, is 8, and the total number of blocks, n, is 255, then CBER $= P_{UE}/(8 \times 255) = P_{UE} \times 4.9 \times 10^{-4}$.

As the encoding and decoding process takes place on a per-block basis it is not surprising that the total latency resulting from encoding and decoding is proportional to the block length n, being typically on the order of 2 to 3 times the block length in time. Thus, for a transmission link where the bit rate is 10 Gb/s and hence each bit occupies 10^{-4} µs, coding is RS(255,235), and each symbol is 8 bits long, then the block length in time is $255 \times 8 \times 10^{-4} = 0.2$ µs, and the total induced latency therefore likely to be on the order of 0.4–0.6 µs.

5.2.6 *LDPC and RS Codes in Wireless Transport*

For the same code rate, LDPC codes provide between about a 2 and 4 dB increase in coding gain compared to RS codes, this increase being a function of code rate, codeword length, and the number LDPC decoding iterations. In modern wireless transport systems, both LDPC and RS FEC codes are usually supported. Because of its superior coding gain performance when block lengths are large, LDPC is normally the preferred choice for backhaul and midhaul applications where the latency requirements are somewhat relaxed (Sect 1.4). For fronthaul application, however, where very low latency is a requirement (Sect. 1.4), RS is a considered option as it can be configured for lower latency relative to LDPC.

As the goal is to achieve improved BER performance via FEC with minimal overhead added, typically no more than about 10% FEC overhead is added in wireless transport links.

5.2.7 *Polar Codes*

Polar codes are block codes and were invented by Arikan and disclosed in 2009 [8]. They are the first error-correcting codes that are theoretically able to achieve the capacity of a *binary discrete memoryless channel*. By memoryless channel, we mean one where the output signal at a time t is only determined by the input signal at time t and consequently not dependent on the signal transmitted before or after t. By capacity, we mean capacity as defined by Shannon [8], who showed that it was theoretically possible for a communication system to transmit information with an arbitrarily small probability of error if the information rate R is less than or equal to a rate C, the *channel capacity*, where, for a channel in which the noise N is bandlimited Gaussian, B is the channel bandwidth, and S is the signal power, C is given by the Shannon Hartley theorem [9]:

$$C = B \log_2\left(1 + \frac{S}{N}\right) \text{ bits/ sec} \tag{5.9}$$

By permitting the achievement of channel capacity, polar codes, in theory, permit the transmission of bits at rate C, the highest rate possible with negligible error. In practice, C is not attainable as it requires unreasonably large block lengths. This disadvantage is offset, however, by the fact that encoding and decoding operations can be performed with relatively low complexity in a deterministic recursive fashion.

5.2.7.1 Channel Polarization

Polar codes are able to approach channel capacity by employing *channel polarization*, hence its nomenclature. With channel polarization, channels are constructed to

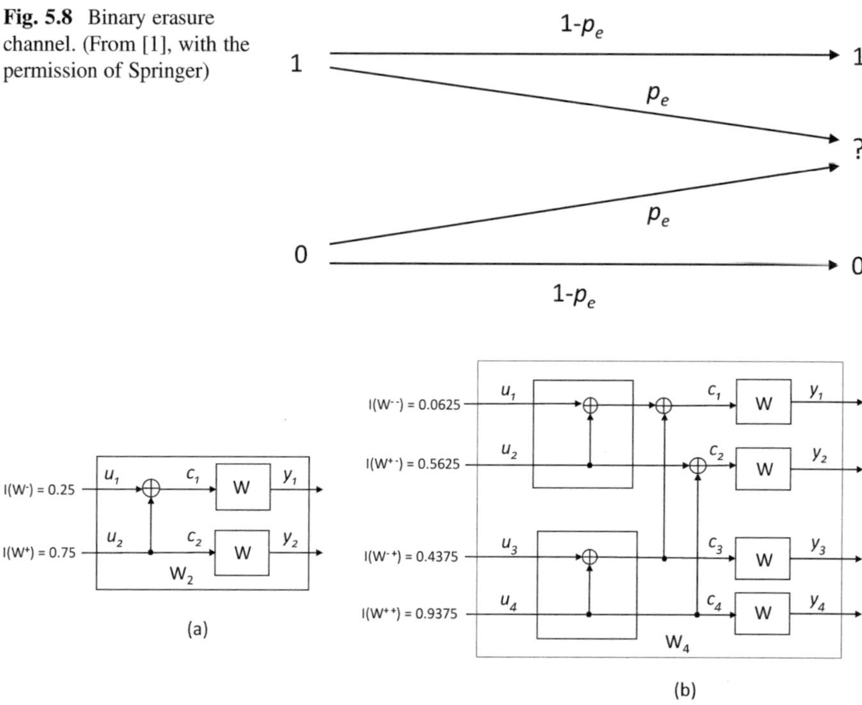

Fig. 5.8 Binary erasure channel. (From [1], with the permission of Springer)

Fig. 5.9 (**a**) Two-channel polarizing combiner. (**b**) Four-channel polarizing combiner. (From [1], with the permission of Springer)

be mostly of high capacity and low capacity. As more and more such channels are constructed recursively, the high-capacity channels get more so and the low-capacity ones get more so. The number of polarized channels used is a function of the code length. Thus, as code length increases, the number of polarized channels and hence polarization increases. A key to polar coding, as we will see below, is that the high-capacity channels are used to transmit the information bits, and the low-capacity ones used to transmit "frozen" bits, normally set to zero.

To aid in describing channel polarization we first introduce, for those unfamiliar with it, the *binary erasure channel* (BEC). The BEC can transmit at any one time only one of two symbols, a 0 or a 1. A model of the BEC is shown in Fig. 5.8. When a bit is inputted to the channel, the channel outputs either the sent bit or a message that it was not received, i.e., erased (erasure symbol is given by ?). If the probability of erasure is p_e say, then the probability of the bit being outputted is $1 - p_e$. Also, it can be shown that the capacity of a BEC is $1 - p_e$.

To understand the polarization effect, we examine Fig. 5.9a, which shows a two-input, two-output channel combiner, employing two BECs, W_1 and W_1, each with capacity $C(W_1)$. These two BECs are combined with the aid of a modulo-2 adder to form the compound channel W_2 with a total capacity of $2C(W_1)$. It can be shown that W_2 can be treated as being *split* into two channels W^+ and W^-, with U_1

being the input to W^- and U_2 being the input to W^+ [9]. With this split structure it can be further shown that the capacity of W^+ is equal to $2C(W_1) - C(W_1)^2$, and the capacity of W^- equal to $C(W_1)^2$ [10]. Thus, though the capacity of each of these new channels is different, the total capacity of the system is preserved. If each W_1 were to have an erasure probability of 0.5, then they would each have a capacity $C(W_1)$ of 0.5, and the capacity of W^+ would be 0.75 and that of W^- would be 0.25. We thus see that under this scenario, the channels have started to polarize. The key to polar coding is that as the number of channels increases the degree of polarization increases. Figure 5.9b shows a four-channel combiner labeled W_4, created by combining two W_2 compound channels, and having the individual channel capacities as shown. As this combining process is repeated the capacity of more and more channels migrate towards either one or zero. Importantly, the polarizing effect works not only for a set of BECs, but also for AWGN channels, where polarization not only addresses capacity, but also BER. However, determining the reliability order of AWGN channels, and hence what channels to assign to information bits, is more complex than for BECs.

   An important feature of polar codes compared to other FEC codes is that they have been shown analytically to not suffer from an error floor.

5.2.7.2   Encoding of Polar Codes

To create a polar encoder, we must know the code block length n to be transmitted, where n must be a power of two, and the number of information bits k per code block. The number of non-information bits, n-k, are referred to, as mentioned above, as frozen bits, and are normally set to 0. The encoder consists of the compounded polarizing encoder W_N. Given the calculated capacities of the individual channels, the k bits are assigned to the channels with the highest capacities and hence the lowest probabilities of error. Figure 5.10 shows a polar encoder for $n = 8$, rate

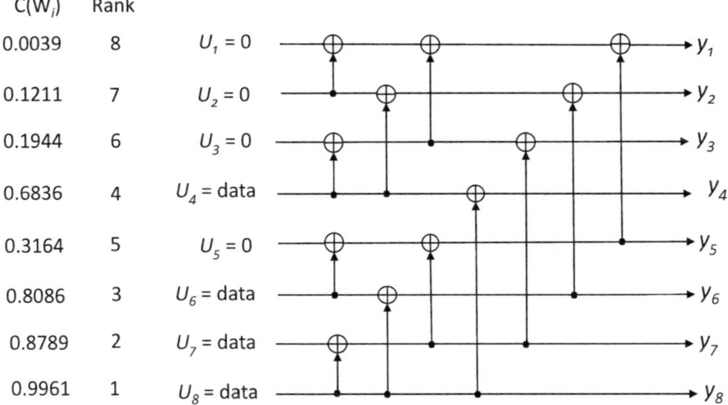

Fig. 5.10 Eight-channel polar encoder. (From [1], with the permission of Springer)

$R = \frac{1}{2}$, and hence $k = 4$ and based on transmission over BEC channels of the probability of erasure of 1/2. Observe that the information bits are assigned to the four channels with the highest capacities.

5.2.7.3 Decoding of Polar Codes

The method normally used to decode polar codes is called *successive cancellation* (SC) decoding [11], a method that is effective enough to achieve capacity at infinite code length. Here, the decoder makes hard bit decisions one at a time, using as the inputs to its computation both the soft information received from the channel in the form of LLRs, as well as the hard decisions made on the previously decoded bits. The algorithm used to determine the value of the bit being decoded is quite complex and involves many LLR computations. If a bit is frozen, then it sets its value to 0. To see how this works, let's consider the decoding of the codes produced by the encoder shown in Fig. 5.10. The decoding proceeds as follows:

Stage 1. Decode U_1. Frozen bit. Hence, $U_1 = 0$.
Stage 2. Decode U_2. Frozen bit. Hence, $U_2 = 0$.
Stage 3. Decode U_3. Frozen bit. Hence, $U_3 = 0$.
Stage 4. Decode U_4. Information bit: Use Y_1 through Y_8, and U_1 through U_3 to decode.
Stage 5. Decode U_5. Frozen bit. Hence, $U_5 = 0$.
Stage 6. Decode U_6. Information bit: Use Y_1 through Y_8, and U_1 through U_5 to decode.
Stage 7. Decode U_7. Information bit: Use Y_1 through Y_8, and U_1 through U_6 to decode.
Stage 8. Decode U_8. Information bit: Use Y_1 through Y_8, and U_1 through U_7 to decode.

An example of a tree representation of SC decoding for a codeword produced by the encoder of Fig. 5.10 is shown in Fig. 5.11. For simplicity purposes, imaginary computed likelihood values (LLR values in real systems) of individual bits being either 1 or 0 are shown beside each node, and the associated bit decision shown next to the preceding tree branch. The path computed by the algorithm is shown in solid lines and rejected paths shown in broken lines. The stages shown are for the stages outlined above. For stages 1, 2, and 3, the decoded decisions are all 0 as these are frozen bits. At stage 4, the likelihood of 1 is higher than that of 0 so 1 is chosen, and the path to the left is terminated. Stage 5 is the decoding of a frozen bit hence the decision is zero. Stages 6, 7, and 8 are decoded as shown, leading to an output sequence of 0 0 0 1 0 0 0 1 and hence an information sequence of 1 0 0 1 (bits number 4, 6, 7 and 8). In general, the decision made on any bit is influenced by all previous bit decisions. If there is an incorrect bit decision, it cannot be corrected later, and thus can result in a cascade of errors in subsequent bits.

Though easy to implement, a concern with SC decoding is its relatively high latency, resulting primarily from the fact the information bits are decoded one by

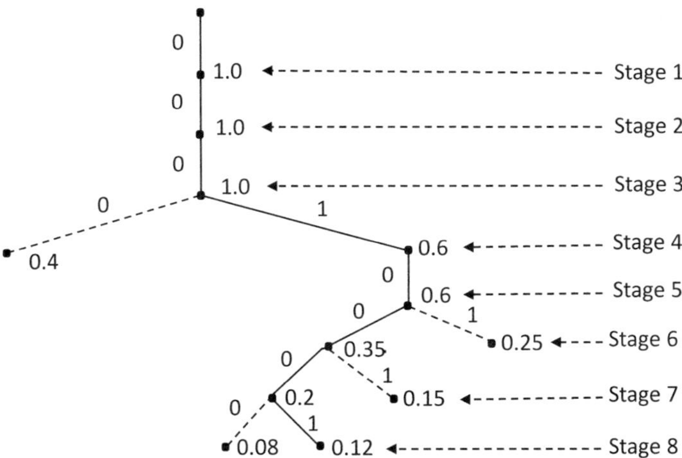

Fig. 5.11 SC decoding of a polar code encoded by the encoder of Fig. 5.10. (From [1], with the permission of Springer)

one. Further, the performance of practical finite length polar codes with SC decoding is noticeably worse than other competitive FEC codes [11]. This latter problem can be addressed with *successive cancellation list* (SCL) *decoding* [10], which substantially improves the BLER performance of SC.

SCL strives to overcomes the limitation of premature decisions taken by SC decoding by employing a list of possible bit sequences, of length L, as it moves from one decoding stage to the next. The list is temporarily doubled at the beginning of each decoding stage, the likelihood of all 2 L paths compared, and only the L paths with the highest likelihoods retained and considered at the next decoding stage. At the end of the process, the SCL outputs the sequence with the highest likelihood. By utilizing a list, the decoder forestalls an early decision which may be incorrect.

An example of a tree representation of SCL decoding for a codeword produced by the encoder of Fig. 5.10 is shown in Fig. 5.12 for a list size of L = 2. As was done for the SC decoder, imaginary computed likelihood values of individual bits being either 1 or 0 are shown beside each node, and the associated bit decision shown next to the preceding tree branch. The path computed by the algorithm is shown in solid lines and rejected paths shown in broken lines. For stages 1, 2, and 3, the decoded decisions are all 0 as these are frozen bits. At stage 4, though the likelihood of 1 is higher than that of 0, the decoder does not make a decision, but keeps both paths under consideration. Stage 5 is the decoding of a frozen bit hence the decision is zero. At stage 6, the encoder computes the likelihood of all four possible paths going forward. However, as the list size is 2, it chooses the two paths with the highest likelihoods and discards the other two. This process of pruning is repeated at stage 7, and at stage 8, it ends up with four possible paths. Here, it chooses the path of

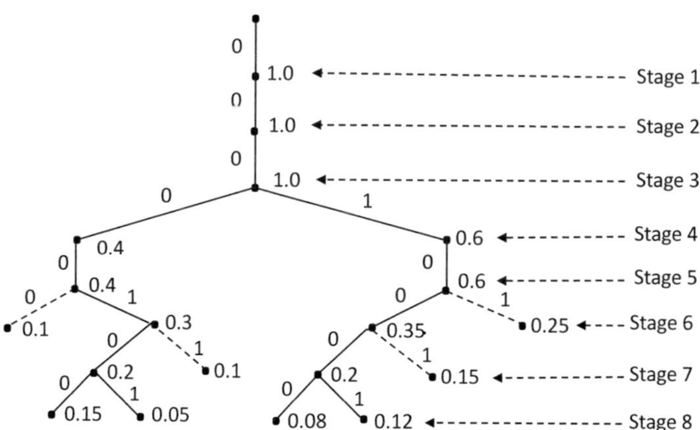

Fig. 5.12 SCL decoding of a polar code encoded by the encoder of Fig. 6.8. (From [1], with the permission of Springer)

highest likelihood, the one ending with a likelihood of 0.15. This leads to an output sequence of 0 0 0 0 0 1 0 0 and hence an information sequence of 0 1 0 0. It will be observed that, with the additional likelihoods given relative to SC decoding tree of Fig. 6.10, the SCL decoded information sequence is quite different from the SC decoded one.

To further improve BLER performance, polar codes can employ *CRC-aided SCL* (CA-SCL) [11]. It was found that when errors occurred with SCL decoding, the correct sequences were usually in the final L-sized list, but that it was not necessarily the codeword with the highest likelihood and hence not necessarily selected in the last step of the decoder. With a CRC added to the codeword, however, the codeword in the list that passes the CRC test can be declared the correct codeword. CA-SCL improves the BLER performance of SCL polar codes, but at the expense of a slight decrease in the code rate. A major concern of SC decoding is large latency. It has been found that this problem can be alleviated by distributing the CRC information bits within the information bits rather than at the end of the information bits.

5.3 Block Interleaving

As we saw above, RS codes reduce the negative impact of error bursts. It is possible, however, to also reduce the negative impact of error bursts by a technique that doesn't involve FEC coding. This technique is called *block interleaving*. A simple example will illustrate how to block interleaving addresses error bursts in a simple block code.

Block A	Block B	Block C	Block D	Block E
A1 A2 A3 A4	B1 B2 B3 B4	C1 C2 C3 C4	D1 D2 D3 D4	E1 E2 E3 E4

(a) Original Codewords

X X   X X X   Error Burst

A1 B1 C1 D1 E1	A2 B2 C2 D2 E2	A3 B3 C3 D3 E3	A4 B4 C4 D4 E4

(b) Interleaved Words

X   X   X   X   X

A1 A2 A3 A4	B1 B2 B3 B4	C1 C2 C3 C4	D1 D2 D3 D4	E1 E2 E3 E4

(c) Deinterleaved Words

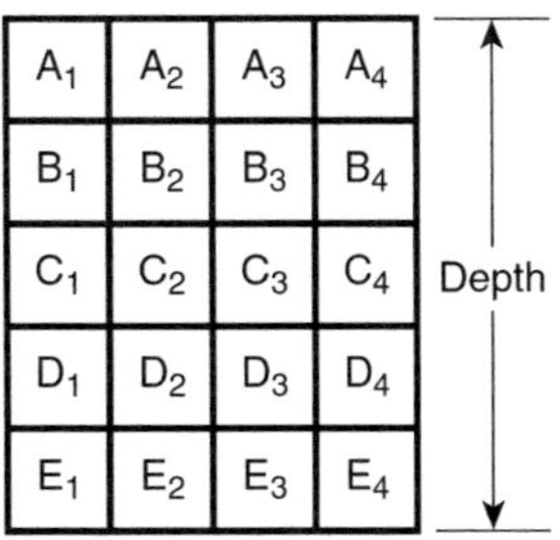

(d) Two-Dimensional Array

Fig. 5.13 Block interleaving/ deinterleaving

Example 5.4: How Block Interleaving Impacts the Decoding of Signals Corrupted with Error Bursts

In this example, an encoder creates the original four-bit codewords shown in
Fig. 5.13a. These codewords are fed to an interleaver that creates the interleaved words shown in Fig. 5.13b. Assume these interleaved words are then
transmitted over a wireless noisy channel. As a result, a burst of five contiguous errors appears on the demodulated interleaved words, as indicated in
Fig. 5.13b. Note, however, that when de-interleaved as shown in Fig. 5.13c,
the five errors are now spread over the five original codewords. Assuming a
decoder that can correct just one bit per codeword, it can, nonetheless, decode
the original five codewords without error. Without interleaving, a burst of five
contiguous errors would have caused errors in two codewords, which would
have been beyond the capability of the decoder to eliminate.

Let N be the length of the codewords. Let the number of coded blocks (codewords) involved in the interleaving process, referred to as the *interleaving depth*, be M. Then the interleaver array is an $M \times N$ one. Thus, our simple interleaver in Example 5.4 has an interleaving depth of five and its array is a 5×4 one. The larger the interleaving depth, the longer the burst of contiguous errors that can be corrected but the greater the delay introduced. A block interleaver consists of a structure that supports a two-dimensional array, of width equal to the codeword length and depth equal to the interleaving depth. For the interleaver of Example 5.4, the data is fed in, row by row, until the array is full, as shown in Fig. 5.13d, then read out column by column, resulting in a permutation of the order of the data. At the receive end of data transmission, the original data sequence is restored by a corresponding de-interleaver. A common designation for an interleaver is π, and that for its corresponding de-interleaver π^{-1}.

Using a block interleaver introduces inherent delay. To see how this occurs, consider the interleaving/deinterleaving shown in Fig. 5.13. At the interleaving end, it can only begin to transmit the first column of the array after the array is full. This way, after the first bit of the first column, i.e., bit A_1 is moved out, a new A_1 bit from a new block following Block E can move in to take its place. If the data input bit rate is r bits/sec, then each bit is of length $1/r$ seconds. As the encoding delay is the time occupied by $M \times N$ bits, then this delay is $(1/_r) \times (M \times N)$ seconds. At the deinterleaving end, a similar delay is incurred.

5.4 Puncturing

Puncturing is the process of discarding some of the bits of an error correction codeword prior to transmission. By employing puncturing, and thus creating punctured codes, it is possible to increase the code rate of a given code. To understand how puncturing works, consider the case where we want to create a rate ¾ code from a rate ½ code. For the rate ½ code, for every three 3-bit input sequence we have a six-bit output. To create the rate ¾ code we simply delete 2 of the 6 output bits, thus giving us 4 output bits for every 3 input ones. The performance of this punctured code is dependent on which bits were deleted. The rate ¾ punctured code can be decoded using the same decoder as required for the original unpunctured rate ½ code. To use the rate ½ decoder, the rate ¾ punctured code is transformed back into a rate ½ structure by simply inserting dummy symbols (1 s or 0 s) into positions where bits were deleted before decoding. The dummy bits result in an impairment of the rate ½ code correcting capability. However, the impaired capability is normally no less than that which would have been achieved had an unpunctured rate 3/4 code been employed in the first place. Punctured codes allow the dynamic selection of code rates based on actual propagation conditions.

5.5 Adaptive Modulation and Coding (AMC)

In original digital microwave links, a single modulation order was employed, and the link fade margin designed so that the desired reliability was achieved in the face of predicted fading due to anticipated atmospheric and other effects. This meant that for perhaps greater than 99% of the time the link was operating tens of decibels above its receiver BER threshold point. This was necessary as there was no way to slow down the transmitted data rate. With packet-based systems, this is not the case. A link can be designed to handle varying rates depending on the varying capacity capability of the link. This is achieved via *adaptive modulation and coding* (AMC). When operating with AMC, a link is designed to operate at an acceptable BER with a high-order modulation scheme, say for example 4096-QAM, and a low overhead coding scheme while the link is in a non-faded condition. As the link starts to fade, however, the modulation order is reduced and if necessary, the coding overhead increased. This improves the receiver sensitivity and hence maintains acceptable BER. This action is not, however, without a penalty. The lower modulation and higher coding overhead necessitate a lowered bit rate in order to maintain the same occupied RF spectrum. As the link comes out of the fade the action is reversed until ultimately operation returns to maximum bit rate. For smooth operation, these changes must take place adaptively and hitlessly.

Consider a link where FEC is employed and the modulation can vary between 4-QAM and 4096-QAM. For these two modulation orders, assuming that the FEC coding gain is the same for both, the difference in receiver sensitivity, i.e., the difference in received signal level required for a BER of 10^{-6}, given the same occupied bandwidth, is approximately 33 dB (see Fig. 4.16 which is for modulation without FEC). This difference in receiver sensitivity is not, however, the only system parameter that changes with change in modulation. As modulation order increases the PAPR of the modulated signal increases. Further, it is also a function of RRC filtering excess bandwidth. The larger the PAPR the greater the power backoff to the transmit power amplifier. The net result is output power that decreases as modulation order increases. For our purposes, let's assume that the output power with 4096-QAM is 5 dB lower than that with 4-QAM. Then the net increase in systems gain in going to 4096-QAM from 4-QAM is 38 dB. If we then had a link with say 40 dB of fade margin when operating with 4-QAM that resulted in an availability of 99.999%, this link could operate when unfaded with a received signal 2 dB above the 10^{-6} BER point with 4096-QAM and thus have the highest bit rate output for a large percentage of the time and outages 0.001% of the time (5 min per year). Bear in mind, however, that with 4-QAM, the bit rate achievable is one-sixth that with 4096-QAM.

5.6   Power Amplifier Linearization Via Predistortion

The transmitter RF power amplifier follows the baseband to RF frequency shifting operation, and its purpose is to provide a high level of output power so that adequate signal level is available to the receiver even with significant fading. The output power of typical wireless transport links varies from about 10 mw to about 5 watts, with maximum offered power decreasing with increasing frequency. The output signal of the RF power amplifier is normally fed via a bandpass filter to the antenna. The bandwidth of this filter is typically wider than the signal spectrum and its purpose is to limit out-of-band radiation as well as to allow duplexing of this signal with an associated incoming signal.

For linear modulation systems with signal states of varying amplitudes, linear amplification is essential to maintain acceptable performance, with higher and higher linearity required as the number of modulation states increases. The effects of nonlinearity on such systems are a nonlinear displacement of signaling states as seen in the constellation diagram and the regeneration of spectral side lobes removed by prior filtering. The displacement of the signaling states degrades the error probability performance while the regenerated spectrum can cause interference to signals in adjacent channels. To avoid these effects the power amplifier must be capable of linearly amplifying all signaling states and thus amplifying the peak signal power. However, high-order linear modulation results in signals where the ratio between peak power and average power can be several dBs. Thus, power amplifiers processing these signals must operate at an average power that is backed off from the peak linear power available by a minimum of the peak to average ratio of the amplified signal.

In a number of wireless transmitter designs, the nonlinearity of the RF power amplifier is counteracted by employing a linearization technique such as *predistortion*. Predistortion works by purposely inserting a nonlinearity into the signal feeding the RF power amplifier that is the complement of the nonlinearity of the RF power amplifier. The preferred modern predistorter is digital and adaptive. In such designs, circuits are added that continuously measure the nonlinearity at the RF amplifier's output and feed this measurement back to the pre-distorter that adjusts its nonlinearity in such a way as to minimize the RF output nonlinearity. By using predistorters, it is typically possible to increase the transmitter output power by low single-digit dBs relative to non-predistorted output power.

Typical input/output power characteristic of a power amplifier is shown in Fig. 5.14a and the required predistorter characteristic for linearization is shown in Fig 5.14b. We note that for the power amplifier, beyond a certain point, as the input power increases the output power deviates for linear and moves towards a saturation level. The aim of the predistorter is to maintain linearity up to the saturation level and thus permit higher output power and increased amplifier efficiency.

A simplified block diagram of a digital predistorted system is shown in Fig. 5.15. Here, the digitized predistorted baseband I and Q signals undergo digital to analog conversion (DAC), low-pass filtering, then modulation, and upconversion. At the output of the power amplifier (PA), a sample of the signal is taken and

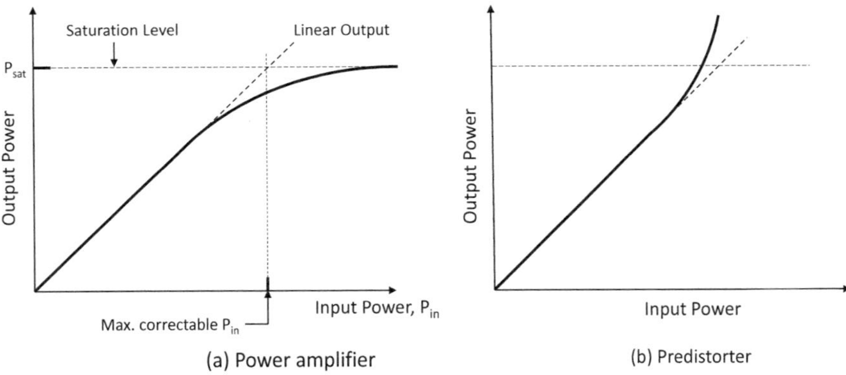

Fig. 5.14 Input/output power characteristic of a power amplifier and its predistorter

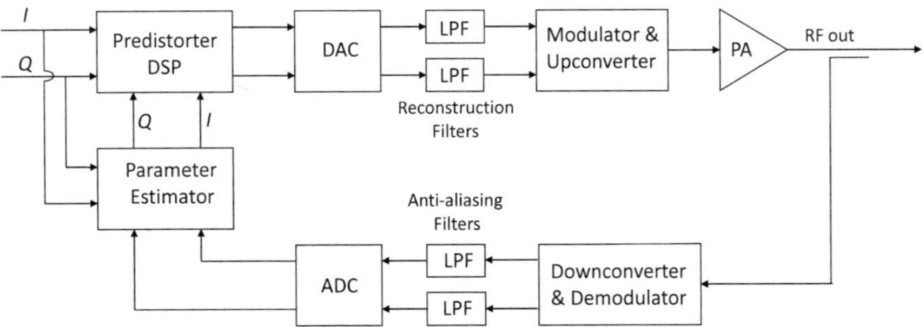

Fig. 5.15 Simplified block diagram of a digital predistortion system

downconverted and demodulated to create I and Q signal streams. These streams undergo low-pass filtering and analog-to-digital conversion (ADC) and are then fed to a parameter estimation unit. This unit is also fed with samples of the digitized I and Q input signals. By comparing the recreated I and Q signals with the original I and Q signals, parameters are computed that feed the predistorter and lead to system linearization.

5.7   Phase Noise Suppression

In an ideal oscillator, the signal created is a pure sine wave and thus with no variation in frequency over time. In a real oscillator, this is not the case. Here, signals suffer disturbances due to thermal noise and device instability. These disturbances have little impact on the signal amplitude close to center frequency but show up mostly as a random variation of the phase and hence frequency about the center position. This phenomenon is referred to as *phase noise* (PN). In point-to-point digital radios, phase noise seen at the receiver is the net result of that due to the transmitter upconversion or direct conversion oscillator and the receiver downconversion or direct conversion oscillator.

The characteristics of an oscillator's phase noise are usually defined via its single-sided *power spectral density* (PSD) . There are a number of models for oscillator PSD and predicted PSD about the oscillator center frequency tends to vary by model. A simple but enduring model is that proposed by Leeson in 1966 [12] for linear feedback free-running oscillators. Leeson's model predicts:

- PSD directly proportional to the noise figure of the oscillator's buffer amplifier and to the oscillator temperature and inversely proportional to signal strength.
- PSD approximately proportional to the square of the oscillator frequency, i.e., increasing by approximately 6 dB for every doubling of the oscillator frequency. Stated another way, when a signal's frequency is multiplied by N, phase noise increases by 20 log (N) dB.
- An initial decrease of about 30 dB per decade up to the point where 1/f noise effects no longer predominate.
- Changing from that point to about 20 dB per decade up to the feedback loop half bandwidth.
- Flattening out thereafter.

A graphical presentation based on this model is shown in Fig. 5.16 [13]. In the model by Demir et al. [14] proposed in 2000, the PSD of a free-running oscillator is modeled as being flat over a range close to its center frequency then falling off at a rate of 20 dB per decade. A more recent model [13] is for phase-locked loop (PLL)

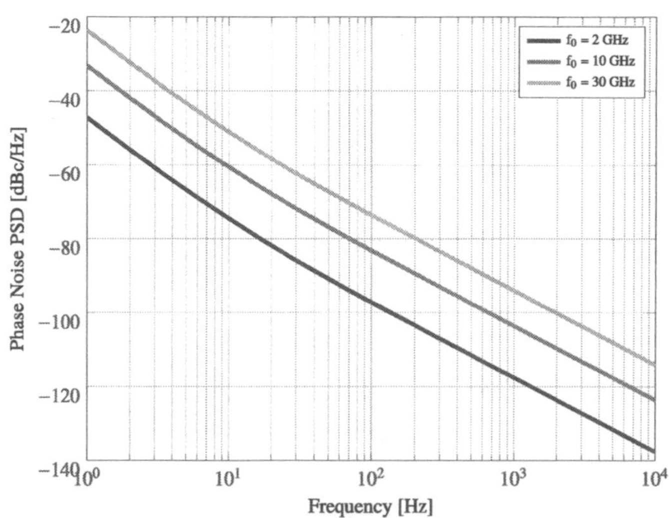

Fig. 5.16 Phase noise PSD as per Leeson. Lower, middle, and upper traces are 2, 10, and 30 GHz, respectively. (From [13] with the permission of Elsevier)

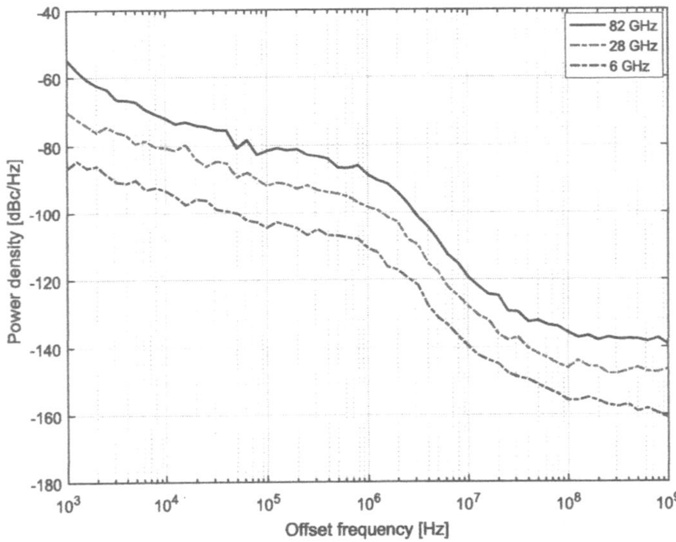

Fig. 5.17 Simulated phase noise PSD. (From [9], with the permission of Elsevier)

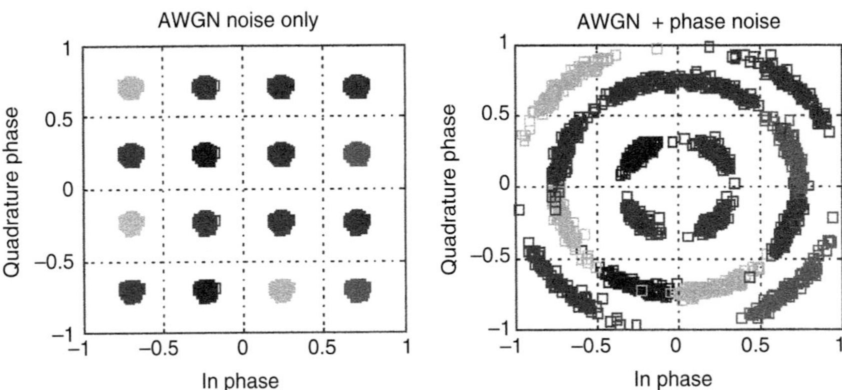

Fig. 5.18 Influence of PN on 16-QAM symbols. (From [15], with the permission of Academic Press)

oscillators which tend to exhibit lower PN than free-running ones but are more costly to implement. This model shows an initial decrease of about 10 dB per decade changing to about 20 dB per decade then leveling off as shown in Fig. 5.17. All three models mentioned above predict that overall, phase noise increases by approximately 6 dB per doubling of the center frequency and this has indeed been found to be the case in practice.

In a single carrier system, phase noise results in a rotation of the constellation. The influence of PN on the received signal in a single carrier 16-QAM system is shown in Fig. 5.18 which is from [15] where the constellation diagram is compared with and without phase noise but including in both diagrams the effect of thermal noise. In the diagram to the right constellation rotation due to phase noise is clear. in We note that, for the same average power, the higher the modulation order, for example, 16-QAM versus QPSK, the more closely spaced are the points in the constellation diagram, and hence the more sensitive the system's BER performance to PN.

In wireless transport links, increased data capacity is afforded by the use of millimeter-wave frequencies where higher channel bandwidths are available. However, as indicated above, phase noise PSD increases with carrier frequency. Thus, PN is a much greater issue at millimeter-wave frequencies than those less than about 20 GHz, and without adequate correction, performance could be significantly degraded. Even though traditional carrier recovery circuits such as covered in Sect. 4.5.2 suppress phase noise, with such circuits it is difficult to achieve adequate suppression when QAM modulation order is very high, when the transmission frequency is very high, or when both of these situations exist simultaneously.

A common way to mitigate against the impact of high PN is to take corrective action in the time domain. Here, pilot-symbols are introduced along with the standard information symbols thus slightly reducing the transport capacity. The symbols are placed between standard high-order QAM multi-valued modulated information symbols in a predefined cycle. For example, one pilot symbol could be placed after 24 information symbols, resulting in a loss of capacity of 4%. Pilot symbols are usually QPSK modulated ones, placed at a level that is equal to the average level of the constellation points of the highest modulation order that is transmitted. To see the advantage of such a system, consider the case where the maximum modulation level is 512-QAM and hence the SNR for a BER of 10^{-6} is 25.9 dB (see Fig. 4.16). At this SNR the BER associated with the QPSK symbols is $\ll 10^{-10}$, thus, unlike the information symbols, wrong decisions about pilot symbols and hence phase error are vanishingly small.

One way the pilot symbol information can be used in the receiver to mitigate against PN is shown in Fig. 5.19. Here, the pilot symbol phase estimate is based on the average of the phases detected on two consecutive pilot symbols, accomplished via the summing of one such symbol with a delayed version of the previous such symbol. This average phase estimate ϕ is then fed into a phase correction generator that outputs a phase correction to the I rail and one to the Q rail, the symbols on these

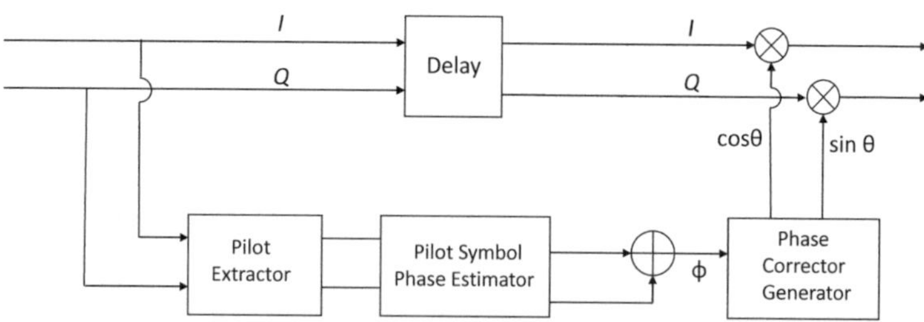

Fig. 5.19 Pilot symbol phase estimator and corrector

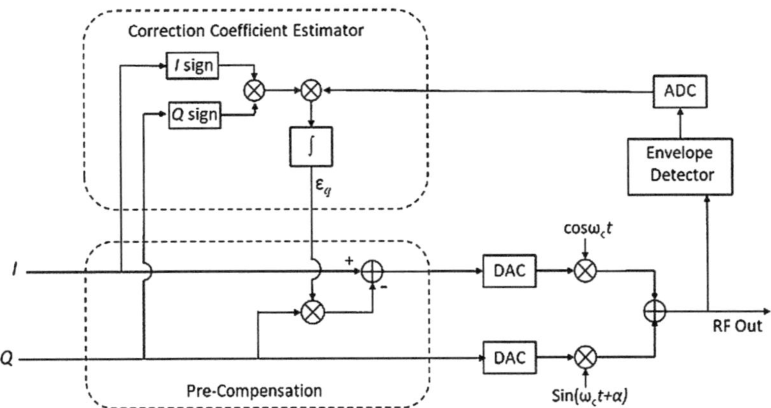

Fig. 5.20 Pilot -based phase noise mitigation in a heterodyne millimeter-wave radio link. (From [17], with the permission of IEEE)

rails being delayed prior to correction so as to be aligned with the correction input. Since we know, a priori, the phase of all undistorted QPSK constellation points (45^0 and multiples thereof), it is relatively easy to compute the phase error. Further, this computed error likely applies equally to all intervening information symbols as phase noise varies slowly with respect to the symbol frequency. In the case of low SNR, one or both of the pilot symbols used for phase estimation could result in flawed estimates leading the errors by all the intervening information symbols. One way to reduce this is risk is to make the phase estimate based on a larger number of pilot symbols. This improvement comes, however, at the cost of increased latency. A system employing this general approach with enhancements, developed by NEC, and referred to as pilot-symbol assisted modulation (PSAM) is outlined in [16].

An alternative way to mitigate against the impact of high phase noise in heterodyne millimeter-wave systems is given in [17]. Here, mitigation is achieved via the addition of an RF pilot tone. As per [17], "the pilot is used as a phase reference to restore an arbitrary signal from phase and frequency impairments at the receiver using only conventional microwave/RF components." Fig. 5.20 outlines the basic concepts of this approach. First, the pilot is added next to the IF modulated data signal at the edge of the modulated bandwidth as shown. The pilot f_p and the IF carrier f_{IF} are synchronized via a phase lock loop to improve frequency stability. We note that since f_{IF} is likely in most cases to be less than 1 GHz, the associated PN will be relatively low. This composite signal is then upconverted, amplified, and transmitted. The upconversion process adds significant PN given that the RF output frequency is a millimeter-wave one. In the receiver, the downconversion process adds more PN. The output of the downconverter feeds the "Phase restore." This is where the key PN mitigation process takes place. The input signal is split and in one arm a bandpass filter (BPF) filters out the pilot and amplifies it and in the other arm, though not shown, a delay equal to that produced by the BPF is added. The signals of the two arms are fed to a mixer with the signal at f_p acting as a new local oscillator. At the mixer output, the phase and frequency impairments incurred via the millimeter-wave up- and downconversion processes are subtracted from the new downconverted signal centered at $f_{IF} - f_p$, and the image signal centered at $f_{IF} + f_p$ carries twice the impairment due to the up- and downconversion processes. At the output of the Phase restore mixer, the image signal is filtered, and coherent detection of the QAM modulated signal centered at $f_{IF} - f_p$ can be accomplished by utilizing a LO frequency of $f_{IF} - f_p$. Alternatively, the signal could be upconverted to the original IF frequency f_{IF} and then coherently detected.

To understand why this process works, let's consider the simple case where the signal carrier is unmodulated:

Let the downconverted carrier at the input to the Phase restore mixer, at a given instant T, be $S_{IF} = A \cos (2\pi f_{IF} T + \theta)$ where θ is the phase noise at time T.
Let the downconverted pilot at the input to the Phase restore mixer, at a given instant T, be $S_p = B \cos (2\pi f_p T + \theta)$ where θ is the phase noise at time T.

Importantly, the acquired phase θ is essentially the same for S_{IF} and S_p as both signals acquired this phase largely from the same up and down local oscillators.

At the mixer output, we now get

$$S_{IF} \times S_p = AB \times \cos (2\pi f_{IF} T + \theta) \times \cos (2\pi f_p T + \theta) \tag{5.10a}$$

$$= \frac{AB}{2} \left[\cos (2\pi (f_{IF} - f_p)T + \cos (2\pi (f_{IF} + f_p + 2\theta)] \tag{5.10b}$$

We note from Eq. (5.10b) that there is no phase component for the signal at $f_{IF} - f_p$ whereas the phase component has doubled for the signal at $f_{IF} + f_p$.

5.8 Quadrature Modulation/Demodulation Imperfections Mitigation

Quadrature modulation/demodulation imperfections distort the constellation diagram resulting in degraded performance, this degradation being proportional to the modulation complexity. The main modulation degradations are:

1. *Quadrature error, α*: Here the phase difference between I and Q modulating and or demodulating oscillator is not $90°$.
2. *I/Q Balance error, δ*: Here the gain on the I rail and the Q rail is not the same.
3. *Residual error, χ*: Also referred to as *LO leakage*, here a continuous component is added to the amplitude of the rail. This can occur when a portion of an oscillator signal feeds through from an input to the output of a mixer and is thus added to the wanted signal.

There are a myriad of techniques that deal with the mitigation of I/Q impairments and we will only touch on a few here. At both the transmitter and receiver, these impairments are usually mitigated in an adaptive manner. When addressed at the transmitter, mitigation action is referred to as pre-compensation as the compensation takes place at the I/Q baseband prior to where the impairment likely occurred. With pre-compensation, a feedback path is typically required, and a large number of such techniques require long training sequences, thus reducing the system capacity. Techniques which do not require training sequences are described by Marchesani [18]. Following is a summary of these techniques.

5.8.1   *Transmitter Quadrature Error Mitigation*

Before reviewing the mitigation process for transmitter quadrature error let's first examine the effect of this error at coherent demodulation. Let's suppose that the signal to be demodulated is $S_q(t)$, where $S_q(t)$ is given by:

$$S_q(t) = I(t) \cos \omega_c t + Q(t) \sin (\omega_c t + \alpha) \tag{5.11}$$

Then the detected I rail is given by $S_q(t) \cos \omega_c t$. Simply mathematical manipulation and ignoring any terms at $2\omega_c t$ that are normally filtered out gives:

$$S_q(t) \cos \omega_c t = \frac{I(t)}{2} + \frac{Q(t)}{2} \sin \alpha \tag{5.12}$$

Similarly, the detected Q rail is given by:

$$S_q(t) \sin \omega_c t = \frac{Q(t)}{2} \cos \alpha \tag{5.13}$$

We thus see that both rails are corrupted leading to an increase in BER.

The squared envelop of $S_q(t)$ as defined in Eq. 5.11, $(EnS_q(t))^2$ can be shown to be given by:

$$\left(EnS_q(t)\right)^2 = [I(t)]^2 + [Q(t)]^2 + 2I(t)Q(t) \sin \alpha \tag{5.14}$$

As shown in [18], if $(EnS_q(t))^2$ is multiplied by $sign(I(t))\, sign\,(Q(t))$, then after integration an error signal ε_q is obtained which is a function of $\sin\alpha$, the sin of the quadrature error α. This signal can be used to cancel the portion of $Q(t)$ in the detected $I(t)$ (See Eq. (5.12)). This is done by multiplying the error signal ε_q by $Q(t)$ and subtracting the result from $I(t)$. Figure 5.21 shows a possible implementation of this adaptive precompensation where baseband activity is assumed to be digitized. We note that this procedure does not compensate for the change in the amplitude of

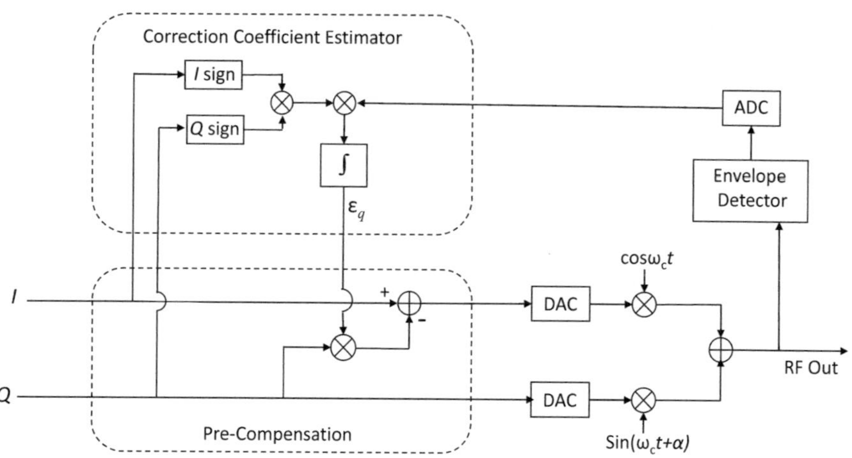

Fig. 5.21 Quadrature error correction

the Q rail resulting from quadrature error (See Eq. (5.13)). Such compensation can be accomplished, however, in balanced error correction.

5.8.2   Transmitter I/Q Balance Error Mitigation

With *I/Q balance error*, also referred to as *I/Q amplitude imbalance*, the modulated signal can be represented as

$$S_{be}(t) = I(t) \cos \omega_c t + (1 + \delta)Q(t) \sin \omega_c t \tag{5.15}$$

Here the squared envelope can be shown to be given by:

$$(EnS_{be}(t))^2 = [I(t)]^2 + [Q(t)]^2 + (2\delta + \delta^2)[Q(t)]^2 \tag{5.16}$$

As shown in [18], if $(EnS_{be}(t))^2$ is multiplied separately by the absolute values of $I(t)$ and $Q(t)$ and the difference of these two products integrated (filtered) then the result is an error signal that is a function of δ and δ^2. As δ is likely to be a small fraction, we can ignore δ^2. This error signal can then be used to compensate for the error balance by adding it to $I(t)$ (See Eq. (5.15)). Figure 5.22 shows a possible implementation of this adaptive precompensation where baseband activity is assumed to be digitized.

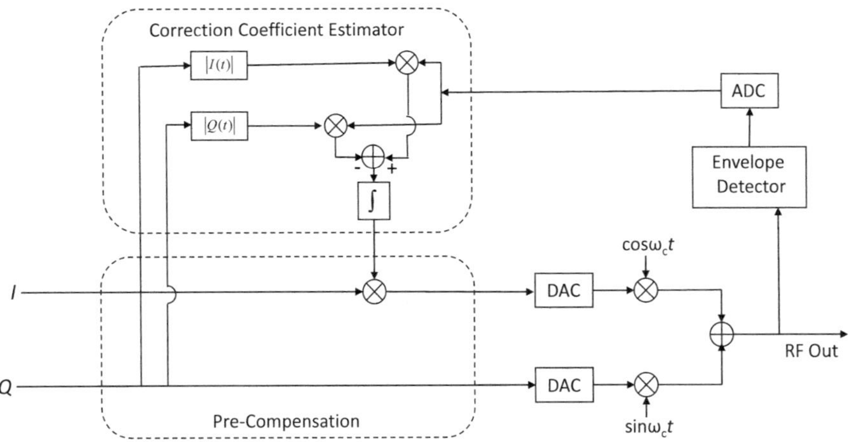

Fig. 5.22 *I/Q* balance error correction

5.8.3   *Transmitter Residual Error Mitigation*

With *residual error*, also referred to as *LO leakage*, the modulated signal can be represented as

$$S_{re}(t) = (I(t) + \chi_I) \cos \omega_c t + \left(Q(t) + \chi_Q\right) \sin \omega_c t \tag{5.17}$$

Here the squared envelope can be shown to be given by:

$$(EnS_{re}(t))^2 = [I(t)]^2 + [Q(t)]^2 + 2I(t)\chi_I + 2Q(t)\chi_Q + \chi_I^2 + \chi_Q^2 \tag{5.18}$$

As shown in [18], if $(EnS_{re}(t))^2$ is multiplied separately by the signs of $I(t)$ and $Q(t)$ and the resulting products filtered, then these filtered outputs are error signals proportional to χ_I and χ_Q, respectively. These error signals can then be subtracted from $I(t)$ and $Q(t)$, respectively, to compensate for the residual error. Figure 5.23 shows a possible implementation of this adaptive precompensation where baseband activity is assumed to be digitized.

5.8.4   *Receiver Quadrature Imperfections Mitigation*

In the transmitter, modulation imperfections were addressed above by detecting the post modulation envelope, digitizing it, and then feeding it back to the digitized baseband circuitry. In the case of the receiver, compensation for the various modem imperfections is typically performed digitally at baseband, i.e., after where the imperfections occurred. Thus, no additional hardware is required to provide

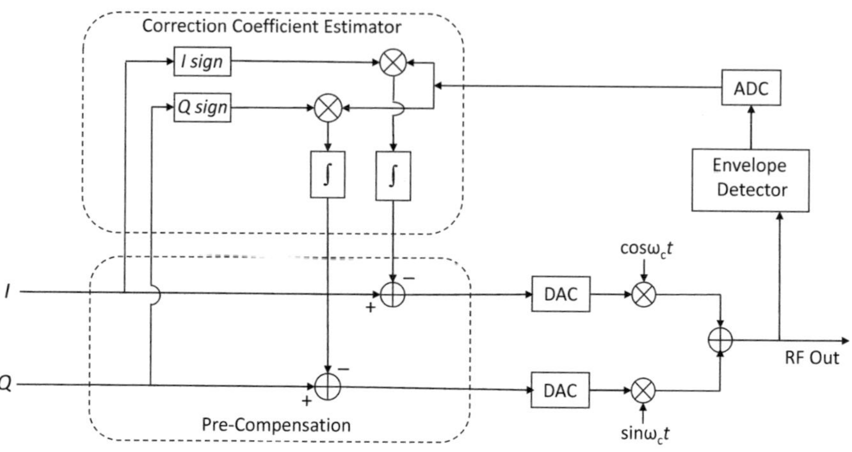

Fig. 5.23 Residual error correction

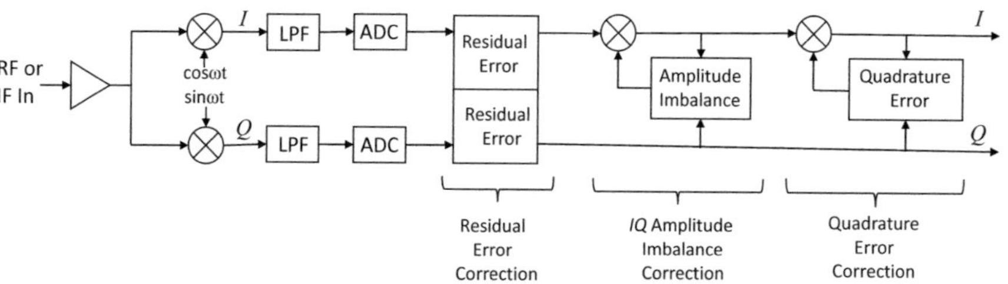

Fig. 5.24 Quadrature modulation imperfections correction in a receiver

feedback between the analog front end and the digital baseband as was the case in the transmitter. As with transmitter mitigation, a number of techniques exist that employ training sequences. However, "blind compensation" techniques, where no training sequence is required, are preferred in wireless links subject to path-induced outages. The compensations are performed right after the analog-to-digital conversion of the demodulated I and Q rails as shown in Fig. 5.24. The correction signals for residual error, I/Q balance error, and quadrature error can all be generated through straightforward computations. These computations first determine the deviation from normal via comparison of the actual I and Q rails with normal, then use these deviations to generate a correction signal that's applied to the appropriate rail as required. We note that in the case of residual error, the error and hence correction of each rail is independent of the other whereas in the cases of amplitude imbalance and quadrature error, the errors are relative ones between the rails, thus, input from both rails is required to determine the error correction. A good review of compensation techniques in both the transmitter and receiver of I/Q imperfections is provided in [19].

5.9 Adaptive Equalization

5.9.1 *Introduction*

As was indicated in Chap. 3, frequency selective fading distorts the amplitude and phase characteristics of the transmitted signal, resulting in *intersymbol interference* (ISI). Further, in all practical systems, there is always some level of ISI due to nonlinearities and the imperfect (non-Nyquist) nature of the combined transmitter and receiver filtering. ISI degrades BER performance and hence, if significant enough, some means must be employed to minimize it. One approach to addresses the effect of frequency selective fading and which operates in the frequency domain is to introduce in the receiver at IF a frequency selective network that has a transfer characteristic that is the inverse of that of the fading channel. However, as the transfer characteristic of the fading channel is time-varying, it is necessary that the equalizing circuit also varies in time, continuously adapting its characteristic in response to the channel. Thus, such a circuit is known as an *adaptive equalizer*. Frequency domain adaptive equalizers were the first types employed in digital radios. Such equalizers were normally a combination of an adaptive slope equalizer in combination with an adaptive notch equalizer. Unfortunately, such equalizers perform poorly in the presence of non-minimum phase fades (Sect. 3.5.3.3) and are relatively costly to implement. It is also possible, however, to compensate adaptively for ISI in the time domain at baseband. Such compensation has the advantage of addressing all ISI acquired prior to decoding and has completely replaced frequency domain equalization.

5.9.2 *Time-Domain Equalization*

5.9.2.1 Introduction

In frequency domain equalization the strategy is to restore as best as possible the modulated signal to its undistorted state. However, the specific cause of BER degradation due to frequency selective fading is, ignoring any decrease in SNR, *inter symbol interference* (ISI), which is a time-domain effect. Equalization in the time domain is therefore the approach that addresses the issue most directly and as a result is the most effective. For a raised cosine filtered received pulse to produce zero ISI, it must have an amplitude versus time characteristic shown in Fig. 4.4, where the amplitude is zero at all sampling instances other than at the one coinciding with peak pulse amplitude. The effect of frequency selective fading is to introduce nonzero sampling responses to the right and left of the pulse. Those responses to the right result in ISI in future pulses, which is referred to as *postcursor* ISI (from past symbols). This ISI dominates during minimum phase fading. Those responses to the left result in ISI in past pulses, which is referred to as *precursor* ISI (from future symbols). This ISI dominates during nonminimum phase fading. The impact of frequency selective fading on ISI is well described by Siller [20]. A highly simplified explanation of how this ISI comes about is as follows: In minimum phase fades, the refracted ray is weaker than the direct ray and is delayed in time relative to it. Intuitively, then, one expects the refracted ray to distort the pulse associated with the direct ray at time periods after its peak. This is indeed the case, resulting in postcursor ISI. In nonminimum phase fading, the weaker ray, arrives at the receiver ahead of the stronger ray. As a result, one now intuitively expects the weaker ray to distort the pulse associated with the stronger ray at time periods before the stronger rays peak. This is indeed the case, resulting now in precursor ISI. Non-minimum phase fading occurs when the refracted ray is delayed relative to the direct ray but has an amplitude that's greater than it. In such a situation, the receiver treats the stronger refracted ray as the direct ray; hence, the true direct ray arrives ahead of the perceived direct ray, leading again to precursor ISI. From the preceding, it is clear that effective time-domain equalization, of signals subject to frequency selective fading, requires the elimination of postcursor and precursor ISI.

5.9.2.2 Adaptive Baseband Equalization Fundamentals

The adaptive form of the baseband *transversal equalizer* (*TVE*), also called the *tapped delay-line equalizer*, is the most common form of *time-domain equalizer* (*TDE*) used on fixed wireless systems. It can be configured in many forms, but before considering some of these, a review of its basic principles is in order. Figure 5.25 shows a block diagram of a *linear feedforward transversal equalizer* in its simplest form. It consists of a delay line with $2\,m + 1$ taps, tapped at intervals of τ_B, where τ_B

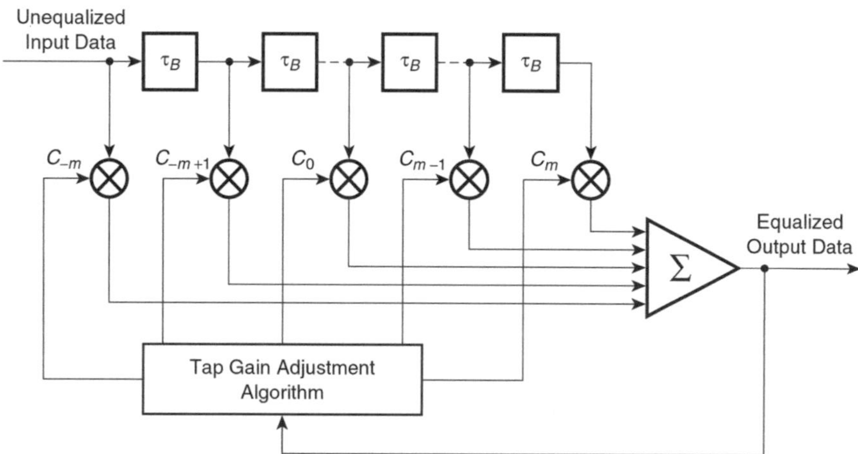

Fig. 5.25 Linear feedforward transversal equalizer with $2\,m + 1$ taps

is the symbol interval of the data stream being equalized. The signal on each tap is weighted by a variable gain factor c, and the weighted outputs are added and sampled at the symbol interval to create the output. The equalizer works by adjusting the tap gains to appropriately weighted versions of preceding and following pulse amplitudes at the prescribed sampling instances, thus canceling interference by them. We note in Fig. 5.25 that if we ignore the feedback from the summer output to the tap again adjustment algorithm, what we have is what is referred to as a *Finite Impulse Response* (FIR) filter. To convey the fundamentals of TVE operation a simple example is presented below.

Example 5.5: Illustration of the Operation of a Baseband Transversal Equalizer

Consider the pulse shown in Fig. 5.26. As a result of transmission impairments, it has a maximum amplitude of 1 V, an amplitude of 0.1 V one sampling interval earlier than the peak, an amplitude of 0.2 V one sampling period later than the peak, but zero amplitude at all other sampling intervals. This pulse, with no nearby adjacent pulses, is applied to a three-tap TVE. When the peak of the pulse is at the first tap of the equalizer, at time t_{-1} say, the output voltage, $v{-}1$, is given by

$$v_{-1} = 1c_{-1} + 0.1c_0 + 0c_{+1} \tag{5.23}$$

One sampling instant later, at time t_0 say, the peak of the pulse is at the second tap, and the output voltage, v_0, is given by

$$v_0 = 0.2c_{-1} + 1c_0 + 0.1c_{+1} \tag{5.24}$$

and one interval later, at time t_{+1} say, the peak of the pulse is now at the third tap, and the output voltage, v_{+1}, is given by

$$v_{+1} = 0c_{-1} + 0.2c_0 + 1c_{+1} \tag{5.25}$$

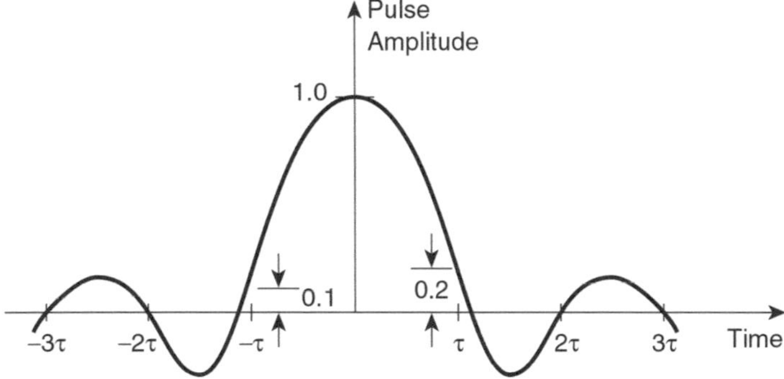

Fig. 5.26 Equalizer input pulse waveform example

Now it's time to stand back and ask just how are we trying to accomplish our goal of zero ISI with this equalizer. A good way to visualize the function of the equalizer is to consider it to be a filter that, in conjunction with the preceding filters, restores the Nyquist criteria for zero ISI. If it succeeds in doing this, the pulse output should have a nominal value of 1 at the sampling instant when it's at its peak, and zero at all other sampling instances, so as not to interfere with other pulses. For our simple equalizer, the desired output pulse sample is v_0 taken at time t_0. Thus, for the equalizer to be 100% successful in removing ISI, v_0 would have to be of value 1 and the outputs at all other sampling instants would have to be zero, assuming that only our one pulse was transmitted. We can prevent it from not interfering with an immediately preceding and immediately following pulse by requiring that $v_{-1} = v_{+1} = 0$. When this is the case, then by Eqs. (5.23) and (5.25) we get

$$c_{-1} = -0.1c_0 \tag{5.26}$$

$$c_{+1} = -0.2c_0 \tag{5.27}$$

and, by substituting Eqs. (5.26) and (5.27) into Eq. (5.24), a peak output voltage v_0 of 0.96 c_0. Hence, for v_0 to be equal to 1, we have

$$c_0 = 1/0.96 \tag{5.28}$$

We have shown that with the tapped signals weighted with the factors found, a single pulse passing through the equalizer should not create ISI in adjacent pulses. To convince ourselves that there is in fact no ISI when there are adjacent pulses, consider the case where a stream of three identical pulses passes through the equalizer. Let t_0 be the time that the center pulse is at the center tap. If the pulses adjacent to it cause it no ISI, then the value of the center pulse, v_0, at time t_0 should be 1.0, just as it was when unaccompanied. To determine v_0, we need to know the sum of the three pulse amplitudes at each tap at time t_0. Labeling the first pulse in as P_1, the second as P_2, and the third as P_3, then at the first tap at time t_0, the amplitudes of P_1, P_2, and P_3, respectively, are 0, 0.2, and 1.0, for a total of 1.2. It can similarly be shown that at the second tap the combined amplitude is 1.3, and at the third 1.1. Multiplying these amplitudes by the determined values c_{-1}, c_0, and c_{+1}, and summing the results, we get, as required,

$$v_0 = (1.2 \times -0.1c_0) + 1.3c_0 + (1.1 \times -0.2c_0) = 0.96c_0 = 1.0 \tag{5.29}$$

Returning now to the case of one pulse passing through the equalizer, we note that because we have no more variables to control the output, we cannot prevent the pulse from causing ISI at sampling instances other than immediately adjacent to itself. Thus, two sampling intervals before the peak of the pulse arrives at the center tap, at time t_{-2} say, the resulting output voltage is

$$v_{-2} = 0.1c_{-1} + 0c_0 + 0c_{+1} = -0.01c_0 \neq 0 \qquad (5.30)$$

and two sampling intervals after the peak of the pulse leaves the center tap, at time t_2 say, the resulting output voltage is

$$v_{+2} = 0c_{-1} + 0c_0 + 0.2c_{+1} = -0.04c_0 \neq 0 \qquad (5.31)$$

Clearly, in Example 5.5, we could have adjusted v_{-2} and v_{+2} to zero had we had two more taps, one on each side of the existing three. Obviously, then, we can further and further decrease the residual ISI by adding more and more taps. In order to force m zero outputs at the sampling instants before the desired pulse sampling instant and thus minimize precursor ISI, and m zero outputs after it, thus minimizing postcursor ISI, the TVE requires $2\,m + 1$ tap points. Because this equalizer forces the output associated with a given pulse to zero at $2\,m$ sampling times, it is called a *zero-forcing* (*ZF*) equalizer. Determining the $2\,m + 1$ weighted values c requires the solution to $2\,m + 1$ linear simultaneous equations. The more dispersed the pulse signal, the more the number of taps that are required to force the ISI to a negligible level. Fortunately, because the time dispersion of pulse signals caused by multipath fading is only over a relatively small number of symbol cycles, equalizers with a manageable number of taps, usually no more than about 20, are very effective in reducing ISI to insignificant levels.

The zero-forcing approach neglects the effect of noise. Further, such an approach is only guaranteed to minimize the worst ISI case if the peak pulse distortion before equalization is less than 100% (i.e., if the data eye pattern at the equalizer input is open). For better performance in the presence of noise one of two well-known adaptation algorithm types is usually employed, these being *least mean square* (LMS) algorithms and *recursive least square* (RLS) algorithms.

With LMS algorithms the tap gains are adjusted to minimize the mean square error at the output of the equalizer, this error being the sum of squares of all the ISI terms, including that at the desired pulse, plus the noise power. These algorithms adapt what is referred to as a stochastic gradient descent method, commencing at an arbitrary point and moving in an iterative fashion in small steps towards the optimal point. An equalizer employing an LMS algorithm maximizes the signal to distortion ratio at its output within the constraints of the number of taps and delay. Such equalizers are relatively easy to design while providing effective performance.

With RLS algorithms, which are somewhat similar to LMS algorithms, the tap gains are adjusted to minimize a weighted linear least square cost function related to the input signals. Compared to LMS, RLS provides faster convergency and has a lower mean squared error, but this at the expense of higher computational complexity and potentially poorer tracking performance as the desired filter response changes with time.

The fundamentals of adaptive equalization with a transversal equalizer employing the ZF, LMS, and RLS algorithms is well presented by Proakis [21]. It's a relatively complex affair, and only a cursory overview will be presented here. In wireless

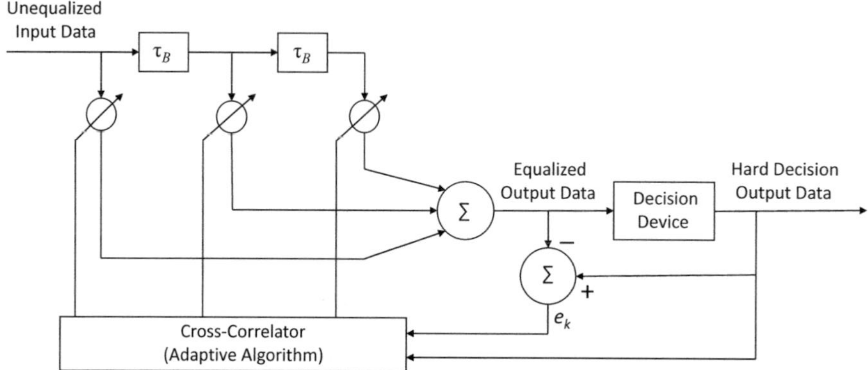

Fig. 5.27 Three-tap decision-directed ZF adaptive transversal equalizer

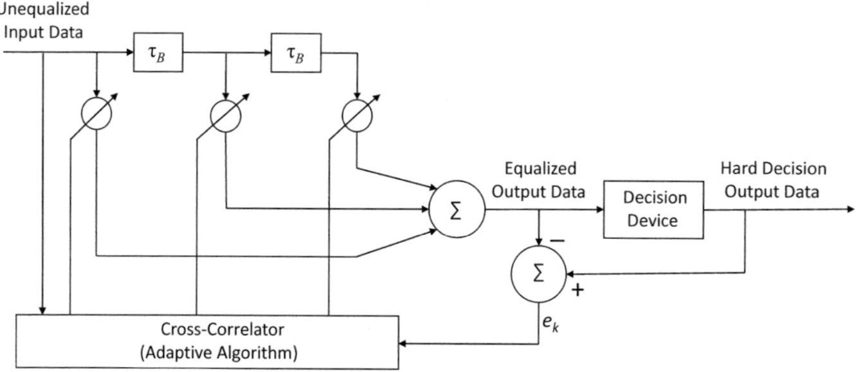

Fig. 5.28 Three-tap decision-directed LMS/RLS adaptive transversal equalizer

systems, for a feedforward equalizer, it is normally achieved by continually adjusting the tap gains during data transmission via an algorithm that is based on the error, or cost function, e_k say, between the signal at the output of the equalizer and the estimate of the transmitted signal made by converting the equalizer output to hard decisions. For the ZF equalizer, this error is correlated with the hard decision data stream to compute estimates of the required tap gains. For the LMS and RLS equalizers, the error is correlated with the input signal to the equalizer. A highly simplified block diagram of a three-tap ZF adaptive equalizer based on this approach is shown in Fig. 5.27, and the equivalent LMS/RLS version is shown in Fig. 5.28. The equalization process is an iterative one, which works to minimize the magnitude of the error signal. Because these equalizers learn by employing their own decisions, the process is called *decision-directed*. For these equalizers to function properly the input symbol sequence must be random; therefore, scrambling is required at the transmitter. As this is always done in fixed wireless systems for other reasons, it does not present a problem.

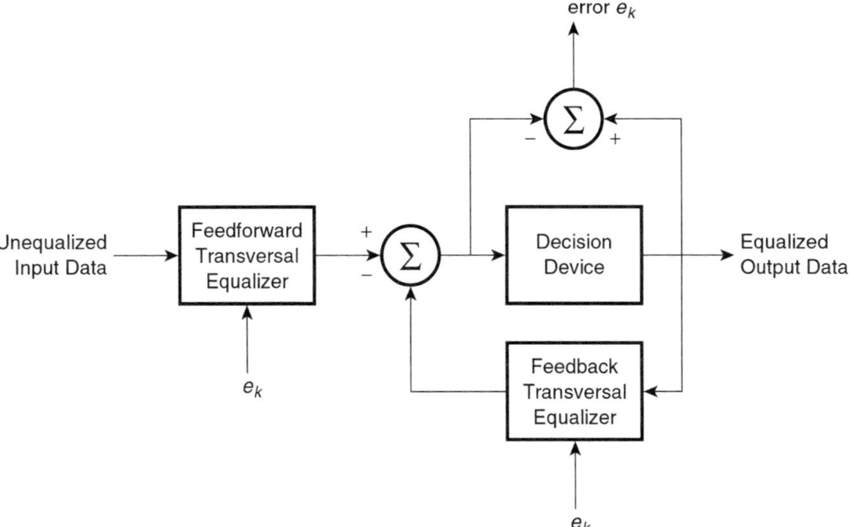

Fig. 5.29 Decision feedback equalizer

A more complex but much more effective version of the adaptive TVE is a nonlinear one, which consists of a feedforward section plus a decision feedback section. Such an equalizer is normally referred to as a *decision feedback equalizer* (*DFE*), with the feedforward component being understood. Note that decision feedback is not to be confused with decision-directed. They have different meanings as will become obvious later. A simplified block diagram of the DFE equalizer is shown in Fig. 5.29. The feedforward section is as discussed previously. The key feature of this equalizer is that, via the feedback section, it uses previously detected symbols (decisions) to eliminate the ISI on pulses that have just been equalized by the feedforward section. The ISI being removed was caused by the trailing edges (in time) of previous pulses. Thus, the distortion on the pulses exiting the feedforward section that was caused by previous pulses (i.e., postcursor interference) is subtracted. Since the feedback section eliminates postcursor ISI, the feedforward section is configured to attempt to compensate only for precursor ISI. Adaptive equalization can be achieved via either the ZF, LMS, or RLS algorithm, with the error sample, e_k, as shown in Fig. 5.29, being inputted to both the feedforward and feedback cross correlators.

Recall that postcursor ISI dominates during minimum phase fading. Because postcursor ISI is canceled in the feedback section, and since this section operates on noiseless quantized levels, its output is free of channel noise. As a result, effective equalization is possible even for minimum phase fades that produce an infinite in-band notch. In nonminimum phase fading, however, where precursor ISI dominates, the behavior of the filter is primarily controlled by the feedforward section. As this section is susceptible to noise, equalization effectiveness is degraded by large in-band notch depth.

5.9.2.3   QAM Adaptive Baseband Equalization

In an ideal QAM demodulator, as described in Chap. 4, the in-phase data stream at the low pass filter output is independent of the modulation on the quadrature phase and vice versa. However, when the transmitted signal is subjected to frequency selective fading, this is no longer the case, with each output rail being contaminated with information from the other. This effect is called *quadrature crosstalk*. To understand this effect, consider a QAM transmission system that is ideal in every respect, except that it is experiencing frequency selective fading. During this fading assume a direct ray and a refracted ray, with the refracted ray having a received amplitude b relative to the direct ray and delayed in time by τ seconds relative to the direct ray. Assume that the direct ray, $r_d(t)$ at the demodulator input, is given by

$$r_d(t) = m_i(t) \cos 2\pi f_c t + m_q \sin 2\pi f_c t \tag{5.32}$$

where $m_i(t)$ is the transmitter baseband in-phase signal modulating signal, $m_q(t)$ is the transmitter baseband quadrature signal modulating signal, and f_c is the carrier frequency.

Then the refracted ray, $r_r(t)$, at the demodulator input, is given by

$$\begin{aligned} r_r(t) &= b r_d(t-\tau) \\ &= b m_i(t-\tau) \cos 2\pi f_c(t-\tau) + b m_q(t-\tau) \sin 2\pi f_c(t-\tau) \end{aligned} \tag{5.33}$$

and the total signal at the demodulator input, $r_I(t)$, is

$$r_I(t) = r_d(t) + r_r(t) \tag{5.34}$$

On the in-phase side of the demodulator, $r_I(t)$ is multiplied by $\cos 2\pi f_c t$, resulting in an output, $r_i(t)$ say. Substituting Eqs. (5.32) and (5.33) into Eq. (5.34) and carrying out this multiplication, using standard trigonometric identities and neglecting the components of the output centered at $2f_c$, we get

$$r_i(t) = \frac{1}{2} m_i(t) + \frac{1}{2} b \cdot m_i(t-\tau) \cos 2\pi f_c \tau - \frac{1}{2} b \cdot m_q(t-\tau) \sin 2\pi f_c \tau \tag{5.35}$$

On the quadrature side of the demodulator, $r_I(t)$ is multiplied by $\sin 2\pi f_c t$, resulting in an output, $r_q(t)$ say. Computing $r_q(t)$ via a similar exercise to that carried out to determine $r_i(t)$ leads to

$$r_q(t) = \frac{1}{2} m_q(t) + \frac{1}{2} b \cdot m_q(t-\tau) \cos 2\pi f_c \tau + \frac{1}{2} b \cdot m_i(t-\tau) \sin 2\pi f_c \tau \tag{5.36}$$

In Eq. (5.35), the first term on the right side is clearly the desired output. The second term is a response due to the delay of the in-phase modulated signal and is

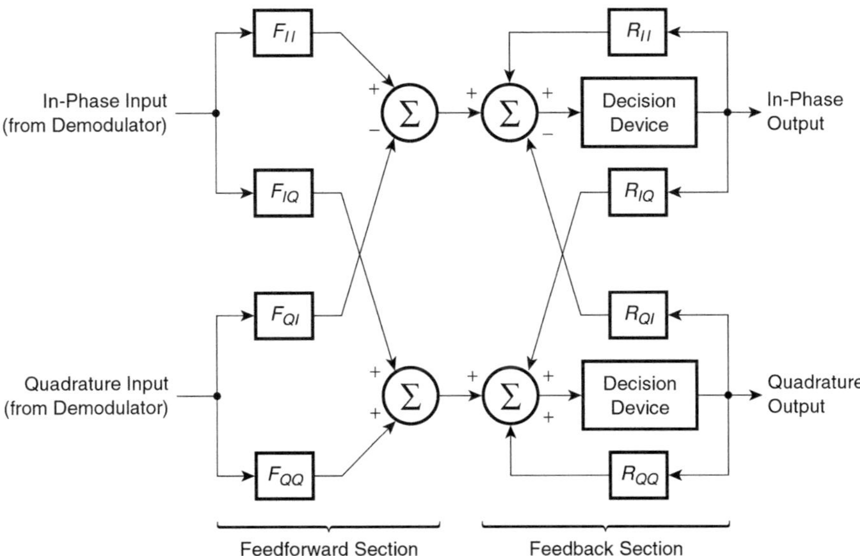

Fig. 5.30 A decision feedback equalizer for QAM systems

referred to as an "in-rail" response. The third term is a "crosstalk" response, being due to the delay of the quadrature modulated signal and is referred to as a "cross-rail response." Similar responses are noted in Eq. (5.36).

As a result of quadrature crosstalk, adaptive transversal equalization of the I and Q data streams of a QAM system cannot be done independently of each other. Figure 5.30 shows the block diagram of a DFE for a QAM system. As can be seen, each input stream drives two feedforward equalizers, one for in-rail equalization, the other for cross-rail equalization. Similarly, each output stream drives two feedback equalizers for in-rail and cross-rail equalization. One may well ask why this doubling of equalizers is necessary. On the I rail, say, wouldn't one feedforward equalizer largely remove all precursor ISI, and one feedback equalizer largely remove all postcursor ISI, regardless of the source? The answer is unfortunately no. The equalizer algorithm only removes ISI resulting from imperfections of a linear nature created within that I channel. As a result, separate equalizers are required to eliminate cross-rail ISI. In Fig. 5.30, note that on the upper signal summers, which handle the in-phase equalization, the inputs from equalizers F_{QI} and R_{QI}, which are driven by quadrature signals, are of negative value, as dictated by the third right-hand term of Eq. (5.35). Similarly, on the lower signal summers, which handle the quadrature equalization, the inputs from equalizers F_{IQ} and R_{IQ}, which are driven by in-phase signals, are of positive value, as dictated by the third right-hand term of Eq. (5.36).

5.9.2.4   Initialization Methods

In the equalization structures just described, it has been tacitly assumed that at the start of the operation the signal at the decision device output is a fairly accurate replica of the original transmitted signal. This condition allows the generation of a mostly valid error signal and hence initial adjustments of the equalizer tap coefficients. This in turn leads to convergence on an ISI free output. In practice, as a result of signal distortion, this is not always the case. It is necessary, therefore, to take action to assure that this condition is quickly met. To accomplish this, one of the two following broad methods is normally applied. In the first method, prior to the transmission of information data, a finite *training sequence*, known to the receiver, is transmitted. In the receiver, a synchronized version of this sequence is generated and used in place of the decision device output signal for error generation. At the end of the sequence, the equalizer switches to the decision device output for error generation. The sequence is made long enough so that the equalizer tap values are adjusted to the point that when regular information data transmission commences the equalizer is fully operative. Such a scheme, though reliable, is not normally used with fixed wireless systems. Because these systems are subject to fading that can interrupt service, then, after every interruption, a message would have to be sent to the transmitter to inject a training sequence. This clearly increases the complexity of the system and is wasteful of throughput capacity.

The second method of initial equalizer alignment, referred to as *blind equalization* [22], is therefore normally used with fixed wireless systems. With this method, no training sequence is transmitted. Instead, during the initial acquisition phase, with knowledge only of the probabilistic and statistical properties of the desired signal, the equalizer adjusts tap values in response to sample statistics instead of in response to sample decisions, as is the case in the normal decision-directed phase. Blind equalization typically employs one of three major algorithms: the *constant-modulus algorithm (CMA)* [23], the *modified constant-modulus algorithm* (MCMA) [24], and the *simplified constant-modulus algorithm* (SCMA) [25]. These algorithms seek to reduce the mean square error to acceptable levels at which point standard ongoing equalization such as decision feedback equalization can take over. In general, they direct the adaptation coefficients towards optimal filter parameters even when the initial error is large. A highly simplified block diagram of an adaptive blind equalizer is shown in Fig. 5.31.

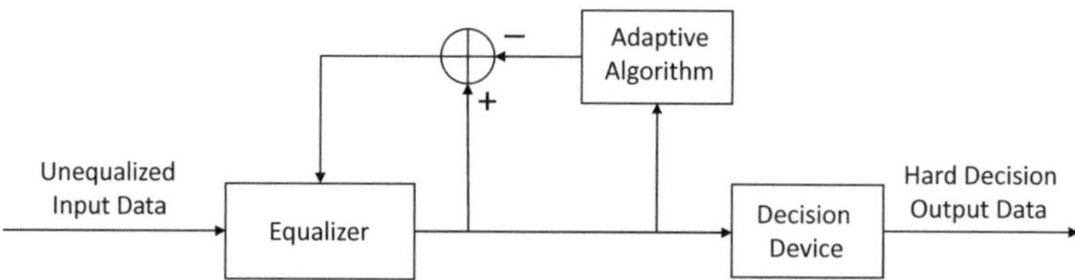

Fig. 5.31 Adaptive blind equalizer

The CMA algorithm is designed primarily for constellations with a constant amplitude such as those generated by phase-shift keying modulation but, as we will see below, can also operate with M-QAM generated constellations where M > 4. The MCMA algorithm improves upon the performance of CMA for M-QAM systems where M > 4. The SCMA, like MCMA, is effective on high-order QAM systems but is less complicated and more flexible than CMA and MCMA from an implementation point of view. A detailed description of these algorithms and equalizers employing them is beyond the scope of this text. However, to get a sense of how they operate, we will briefly review the salient features of the constant-modulus algorithm.

Consider a QAM equalizer where the output of the in-phase adaptive filter for the nth symbol is I_n and the output of the quadrature adaptive filter of the nth symbol is Q_n. Then, the equalizer outputs can be represented as one complex output, Z_n, where $Z_n = I_n + jQ_n$. The CMA equalizes the QAM signal constellation by finding the set of tap values that minimizes the cost function $D^{(p)}$, where $D^{(p)}$ is the expectation (average value) of the square of the difference between the modulus of the equalizer complex output, $|Z_n|$, raised to the power p, and a positive real constant, R_p, that is a function of the constellation structure. Thus,

$$D^{(p)} = E\left(|Z_n|^p - R_p\right)^2 \tag{5.37}$$

For the case of $p = 2$, the algorithm is well documented and relatively easy to implement. As can be seen from Eq. (5.37), the minimization of $D^{(p)}$ is in effect an effort to fit the complex equalizer output to a ring of constant modulus (i.e., magnitude), hence the nomenclature of the algorithm. The function $D^{(p)}$ characterizes the amount of ISI at the equalizer output [23]; thus, its minimization leads to the minimization of ISI. For signals possessing a constant modulus (e.g., 8-PSK) excessive dispersive fading, noise, or interference will significantly vary the modulus of the modulated states, which show up at the equalizer output as a significant variation of Z_n. Thus, it seems intuitively that the minimization of $D^{(p)}$, which forces Z_n back to its original modulus, leads to the reduction of ISI and hence to partial equalization. What is remarkable about the CMA-based equalizer, and certainly not intuitive, however, is that it also partially equalizes signals not possessing a constant modulus, such as M-QAM signals where M > 4.

A key feature of the CMA approach is that, again as can be seen from Eq. (5.37), $D^{(p)}$ is independent of the carrier phase. Thus, its operation is phase invariant and can operate independently of the receiver carrier-tracking loop. However, before switching from the blind mode to the decision-directed mode, separate constellation phase recovery is necessary in order to properly align the constellation.

5.10 Summary

Digital wireless systems employing linear modulation methods are particularly susceptible to the in-band distortions created via multipath fading, this susceptibility increasing as the number of modulation states increases. Systems with high-order modulation, in addition to being highly susceptible to in-band distortion, are also susceptible to their own implementation imperfections. This makes the attainment of error rate performance close to ideal difficult to achieve. A number of highly effective techniques have been developed to address these susceptibilities. As a result, by their application, the transmission of very high data rates is possible with systems employing high-level modulation and thus achieving high levels of spectral efficiency. In this chapter, some of the more important of these techniques that are or may be applied in wireless transport links were reviewed. These techniques included low-density parity-check (LDPC), Reed-Solomon, and polar forward error correction codes, adaptive modulation and coding (AMC), power amplifier linearization, phase noise suppression, quadrature modulation/demodulation imperfection mitigation, and adaptive equalization.

References

1. Morais DH (2020) Key 5G physical layer technologies: enabling mobile and fixed wireless access. Springer, Cham
2. Gallager RG (1963) Low density parity codes. MIT Press, Cambridge, MA
3. Fossorier MPC (2004) Quasi-cyclic low-density parity-check codes from circulant permutation matrices. IEEE Trans Inform Theory 50(8):1788–1793
4. Mackay DJC (1999) Good error-correcting codes based on very sparse matrices. IEEE Trans Inform Theory 45:399–431
5. Burr A (2001) Modulation and coding for wireless communications. Pearson Education, Harlow
6. Gallager RG (1968) Information theory and reliable communication. Wiley, New York
7. Advanced Hardware Architecture (1995) Primer: Reed-Solomon error correction codes (FEC). Pullman, Washington
8. Arikan E (2009) Channel polarization: a method for constructing capacity achieving codes for symmetric binary-input memoryless channels. IEEE Trans Inform Theory 55(7):3051–3073
9. Taub H, Schilling DL (1968) Principles of communication systems, 2nd edn. McGraw Hill, New York
10. Arikan E (2008) Channel polarization: a method for constructing capacity-achieving codes. In: Proceedings of the IEEE International Symposium on Information Theory, Toronto, Canada, July 2008, pp 1173–1177
11. Tal I, Vardy A (2015) List decoding of polar codes. IEEE Trans Inf Theory 61(5):2213–2226
12. Leeson DB (1966) A simple model of feedback oscillator noise Spectrum. Proc IEEE 54(2):329–330
13. Zaidi A (2018) 5G physical layer: principles, models and technology components. Academic Press, London
14. Demir A et al (2002) Phase noise in oscillators: a unifying theory and numerical methods for circuits characterization. IEEE Trans Circuits Syst 1 Fundam Theory Appl 47(5):655–674

15. Dahlman E et al (2021) 5G NR, Second Edition: the next generation wireless access technology. Academic Press, London
16. NEC (2016) Development of a phase noise compensation method for a super multi-level modulation system that achieves the world's highest frequency usage efficiency. NEC Tech J 10(3):83–87
17. Chen J, et al (2014) Experimental demonstration of RF-Pilot-based phase noise mitigation of millimeter-wave systems. In: IEEE 80th vehicular technology conference, Vancouver, Canada
18. Marchesani R (2000) Digital precompensation of imperfections in quadrature modulators. IEEE Trans Commun 48(4):552–556
19. De Witt JJ (2011) Ph.D. Dissertation: modelling, estimation and compensation of imbalances in quadrature transceivers. Stellenbosch University, South Africa
20. Siller C A Jr. (1984) Multipath propagation: its associated countermeasures in digital micro-wave radio. IEEE Communications Magazine, vol 22, no 2, New York
21. Proakis JG (2008) Digital communications: fifth edition. McGraw Hill, New York
22. Garth L, et al (1998) An introduction to blind equalization, TD-7 of ETSI/STS TM6, Madrid, Spain
23. Godard DN (1980) Self-recovering equalization and carrier tracking in two-dimensional data communications systems. IEEE Trans Commun 28(11):1867–1875
24. Oh KN, Chin YO (1995) Modified constant modulus algorithm: blind equalization and carrier phase recovery algorithm. In: Proceedings of the IEEE international conference on communications, vol 1, pp 498–502, Seattle, WA, USA, 18–22 June 1995
25. Lkhlef A, Guennec DL (2007) A simplified constant modulus algorithm for blind recovery of MIMO QAM and PSK signals: a criterion with convergence analysis. Eurasip J Wirel Commun Netw 2007:1–13

Chapter 6
Non-Modulation-Based Capacity Improvement Techniques

6.1 Introduction

In Chap. 4, it was shown that for a given bandwidth radio link capacity can be improved by increasing the modulation order applied. This improvement is achieved, however, at the expense of reduced receiver sensitivity and increased complexity. From a practical point of view, therefore, there is a limit to how much modulation-based capacity improvement is appropriate if alternative improvement methods are available. In Chap. 5, it was shown that with forward error correction coding (FEC) receiver sensitivity can be increased. This increase is achieved, however, at the expense of useful capacity as redundant information is transmitted. In this chapter, we describe additional techniques employed in point-to-point links in order to enhance capacity which, when used along with modulation-based capacity improvement, can lead to extremely high-capacity capability. These techniques are (a) co-channel dual polarization (CCDP) transmission accompanied by cross-polarization interference cancellation (XPIC) in the receiver, (b) line-of-sight multiple-input multiple-output (LoS MIMO) transmission, (c) orbital angular momentum (OAM) multiplexing, and (d) band and carrier aggregation (BCA).

6.2 Co-Channel Dual Polarization (CCDP) Transmission

The use of *co-channel dual polarization* (CCDP) doubles link capacity. With this technique, two independent streams of data are transmitted over the same frequency channel, but with one transmission being vertically polarized and the other horizontally polarized, the transmissions thus are, in theory, orthogonal to one another. The dual modulated signals are typically transmitted on a single dual-polarized antenna and received on a single dual-polarized antenna that separates them and feeds them to two demodulation paths. This approach would work well without any corrective

© The Author(s), under exclusive license to Springer Nature Switzerland AG 2021
D. H. Morais, *5G and Beyond Wireless Transport Technologies*,
https://doi.org/10.1007/978-3-030-74080-1_6

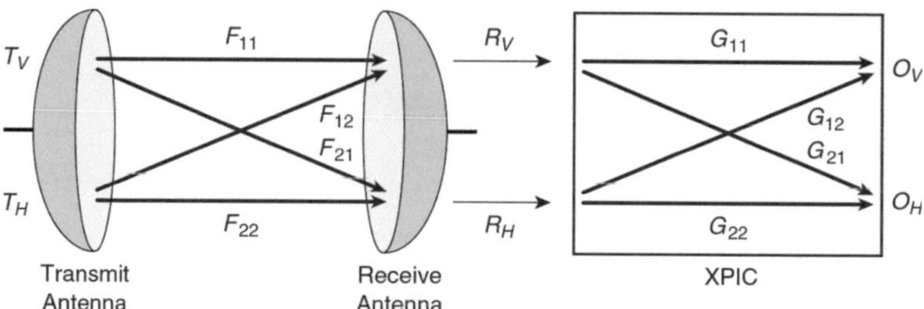

Fig. 6.1 Cross-polarization interference and cancellation model

circuitry as long as the discrimination between the two polarized received signals
was to be so high that what little of one leaks into the other simply appears as
negligible interference. This would be the case with ideal cross-polarization antennas
and close to ideal propagation conditions. In the real world, unfortunately, this is
rarely the case. Antennas are not ideal, propagation can be impaired by factors such
as rainfall and multipath, and processing equipment will likely have imperfections.
This all leads to measurable *cross-polarization interference* (XPI). Thus, to ensure
acceptable operation, it is necessary to apply techniques that minimize XPI. This is
accomplished by the use of an adaptive *cross-polarization interference canceller*
(XPIC) [1].

The principle of operation of an XPIC may be explained with the aid of
Fig. 6.1. T_V and T_H are the transmitted signals, R_V and R_H are the received signals,
and O_V and O_H are the output signals of the XPIC. F_{ij} and G_{ij} are the transfer
functions of the transmission channel and the XPIC, respectively. From the figure, it
follows that

$$R_V = F_{11}T_V + F_{12}T_H \tag{6.1}$$

$$R_H = F_{22}T_H + F_{21}T_V \tag{6.2}$$

and

$$O_V = G_{11}R_V + G_{12}R_H \tag{6.3}$$

$$O_H = G_{22}R_H + G_{21}R_V \tag{6.4}$$

Substituting Eqs. (6.1) and (6.2) into Eqs. (6.3) and (6.4), we get

$$O_V = (G_{11}F_{11} + G_{12}F_{21})T_V + (G_{11}F_{12} + G_{12}F_{22})T_H \tag{6.5}$$

$$O_H = (G_{21}F_{11} + G_{22}F_{21})T_V + (G_{22}F_{22} + G_{21}F_{12})T_H \tag{6.6}$$

XPIC works by making values of G_{ij} such that O_V and O_H are, at all times, good approximations of the transmitted signals T_V and T_H, respectively. Clearly, to accomplish this, it must force $(G_{11}F_{12} + G_{12}F_{22})$ and $(G_{21}F_{11} + G_{22}F_{21})$ toward zero and $(G_{11}F_{11} + G_{12}F_{21})$ and $(G_{21}F_{11} + G_{22}F_{21})$ towards a constant.

The above exercise shows that, mathematically at least, interference cancellation can be accomplished. The challenge, therefore, is to create realizable circuits that can accomplish this cancellation. XPICs can be implemented at RF, IF, or baseband. As with ISI equalizers, the baseband version is the easiest to realize and hence the most common.

In Fig. 6.2, the paths of in-phase (I) and quadrature (Q) symbols streams over a QAM modulated dual-polarized system is shown. As a result of XPI, the I and Q symbol streams at the output of the H channel demodulator, as well as those at the output of the V channel demodulator will suffer from distortion. Thus, to remove all distortion, four separate XPICs must be utilized.

The fundamental structure of a typical baseband XPIC consists of two FIR filter types of primary blocks, namely, a main channel adaptive transversal equalizer (MC-ATE) and a cross-channel adaptive transversal equalizer (XC-ATE). Figure 6.3 shows such a structure for processing the demodulated I stream received via the vertical (V) channel. The MC-ATE (Sect. 5.9.2) compensates for the symbol

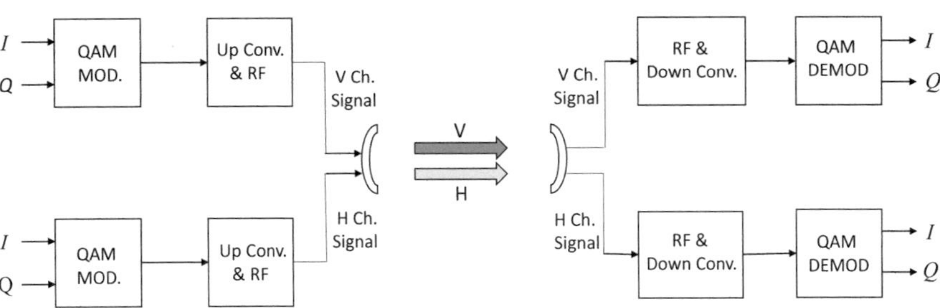

Fig. 6.2 I and Q symbol stream paths over a QAM system

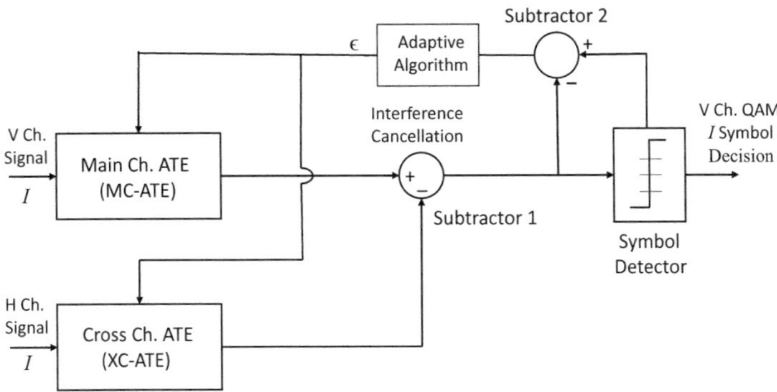

Fig. 6.3 Typical XPIC structure

distortion resulting from the V channel transmission imperfections. This compensation does not remove, however, symbol distortion resulting from the presence of XPI from the H channel at the input to the V channel modem. The XC-ATE compensates for symbol distortion resulting from the H channel transmission imperfections. The output from the XC-ATE is subtracted from that of MC-ATE by Subtractor 1 at such a level, after adaptation, as to result in an output from the subtractor that has minimal distortion due to XPI. The output of the Subtractor 1 is fed to the I stream symbol detector. It is also fed to Subtractor 2 where it is subtracted from the hard output of the detector. The output of this subtractor is fed to the adaptive algorithm whose output, error ϵ, is then fed into both the MC-ATE and the XC-ATE where it drives the tap gain coefficients, resulting in the minimization of distortion due to main channel impairment and XPI respectively. We note that MC-ATE is the realization of G_{11} in Fig. 6.1 and XC-ATE is the realization of G_{12} in Fig. 6.1. For XPIC on all I and Q streams on a QAM cross-polarized system, a structure as shown in Fig. 6.3 must be repeated four times as indicated above.

The structure shown in Fig. 6.3 is one operating in the decision-directed mode where the algorithm used is likely to be the least mean square (LMS) [2] algorithm or the recursive least mean square (RLS) [2] algorithm, both of which were discussed in Sect. 5.9.2.1. However, before operating in this mode, the FIR filter taps need to be initialized. Initialization can be achieved via a training sequence, but this is not the preferred approach with LoS systems. Rather, blind equalization [3] is preferred. Here filter taps are updated based on algorithms such as the Modified Constant-Modulus algorithm (MCMA) [4] and the Simplified Constant-Modulus algorithm (SCMA) [5] which were also discussed in Sect. 5.9.2.1.

In XPICs, such as shown in Fig. 6.3, in order to correctly cancel interference between channels, the phases of the main channel I and Q signals and those of the cross-channel must coincide thus calling for carrier synchronization between the H and V signals. Further, symbols on the H and V channels must be aligned. Circuitry to effect these requirements must therefore be implemented for effective interference cancellation.

We note that a baseband XPIC is normally implemented digitally, thus either the downconverter output is fed to an analog-to-digital converter (ADC) whose digitized output feeds the demodulator, or the demodulator outputs are fed to ADCs and the processing thereafter is all digital.

6.3 Line-of-Sight Multiple-input Multiple-output (LoS MIMO)

6.3.1 Introduction

MIMO technology gained wide acceptance in the non-line-of-sight (NLoS) mobile access environment where its effectiveness is dependent on multipath transmission resulting from the rich signal scattering across the path. As it turns out, however, MIMO can also be applied to line-of-sight (LOS), point-to-point wireless links with no rich scattering across the path. Such links typically employ highly directive antennas which limits the impact of reflected or scattered signals. As with NLoS MIMO, LoS MIMO allows improved capacity for a given frequency allocation, i.e., improved spectral efficiency. It achieves this by creating an artificial transmission multipath, not caused by scattering or reflecting by physical objects, but rather via defined spatial separation of multiple antennas at both the transmit and receiving end. This results in signals at the receiver that appear orthogonal to each other and hence resolvable.

LoS MIMO [6, 7] is a new technology that has found application in mobile network wireless transport links. Compared to links operating in the millimeter-wave range, links operating at frequencies at or below about 12 GHz tend to suffer more atmospheric multipath which degrades the normal operation of LoS MIMO. Further, as we shall see below, links operating in the lower frequency range require large antenna separation that can be difficult to implement. As a result, LoS MIMO links tend to be implemented mainly at frequencies in or close to the millimeter-wave range.

6.3.2   LoS MIMO Fundamentals

LoS MIMO typically utilizes N transmitters and N receivers, all using the same frequency channel. For such systems, capacity is improved by a factor of N. The most common configuration, and the easiest to describe is for the case where N equals two, i.ie, two transmitters and two receivers. We will thus consider such a system which is depicted for clarity in one direction of transmission in Fig. 6.4, though normally such a system is two way. Spatial separation between the two antennas at the transmit end and between the two antennas at the receiving end is denoted by h_1 and h_2 respectively. The distance between transmitter i and receiver j is indicated by d_{ij}. Finally, the distance between the center point of the transmit antennas and the center point of the receive antennas is denoted D. We note that though here the separation between antennas is shown vertically, this separation can also take place horizontally. On towers, vertical separation is normally more easily accommodated, but on a rooftop say, a horizontal separation would likely be more convenient.

The key to LoS MIMO is the different phase shifts along the different propagation paths between the transmit and receive antennas. When these phase shifts satisfy certain conditions, the receivers are able to use signal processing techniques to eliminate interference and recover the originally transmitted signal at a maximum level. The key condition to be met in a 2×2 system is that the path difference between the two paths taken by signals arriving at a given receiver be $\lambda/4$, i.e., a 90^0 difference in phase. With reference to Fig. 6.4 this implies that $d_{21} - d_{11}$ must equal $\lambda/4$ as must $d_{12} - d_{22}$. To see how these differences can lead to interference cancellation we turn to Fig. 6.5 which depicts a vector representation of the signals

Fig. 6.4 General 2×2 LoS MIMO arrangement

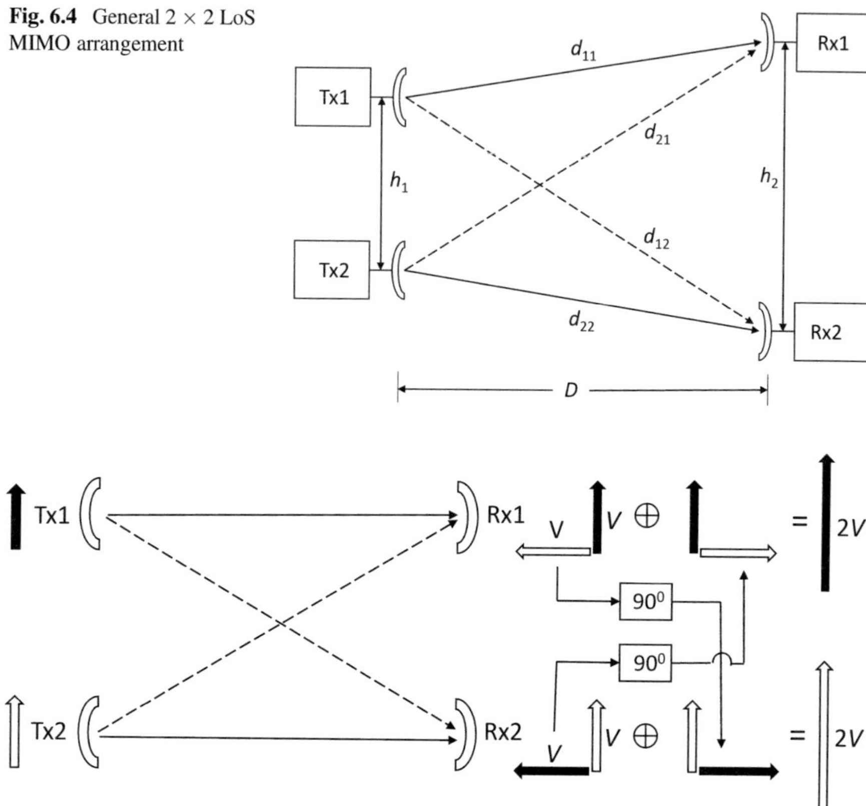

Fig. 6.5 Interference cancellation in a 2×2 LoS MIMO system with optimum antenna spacing

and where the transmitter signals are assumed phase locked, i.e., 0^0 phase difference between them. At Rx1 we show the desired signal from Tx1 vertically, and the interfering signal from Tx2 horizontally. At Rx2 we show the desired signal from Tx2 vertically, and the interfering signal from Tx1 horizontally. What the interference cancelling circuitry accomplishes is to take a copy of the signal received at Rx1, rotate it by 90^0, and add it to the signal received by Rx2. From the figure, we see that by this action, the desired signal at Rx2 is doubled, and crucially, the interfering signal is cancelled. A similar action results in interference cancellation to the signal received at Rx1. As the signals received from the same transmitter are doubled, i.e., added coherently, this leads to a 6 dB increase in relative signal power, whereas the associated noise power is added linearly leading to a 3 dB increase in relative noise power. The net result is a 3 dB increase in SNR relative to a single receive antenna system and hence a 3 dB increase in fade margin. This improvement comes, of course, with a doubling of the total transmit power relative to a single-input single-output (SISO) link. Should the transmit power at each transmitter be halved so that the total power be the same as a SISO link, then the optimal 2×2 LoS MIMO system is seen to support two streams of data with performance equal to the SISO system. It should be noted that though, for the signal analysis above, we choose to have the transmitter signals phase-locked, this is not a necessary condition to achieve full interfering signal cancellation and optimum desired signal addition. No phase relationship is required.

6.3.3   Optimal Antenna Separation

What are the geometric conditions to be met to result in the path difference between the two paths taken by signals arriving at a given receiver be $\lambda/4$ and hence result in optimal performance? To determine this, we refer to Fig. 6.6. from which we can determine that

$$d_{11}^{2} = D^{2} + \left[\frac{h_2 - h_1}{2}\right]^2 \tag{6.7}$$

and

$$d_{21}^{2} = D^{2} + \left[\frac{h_2 + h_1}{2}\right]^2 \tag{6.8}$$

Using the approximation that $\sqrt{p^2 + q^2} \simeq p + \frac{q^2}{2p}$ if $p \gg q$, then we can show that

$$d_{11} \simeq D + \left[\frac{h_2^{2} + h_1^{2} - 2h_1 h_2}{8D}\right] \tag{6.9}$$

and

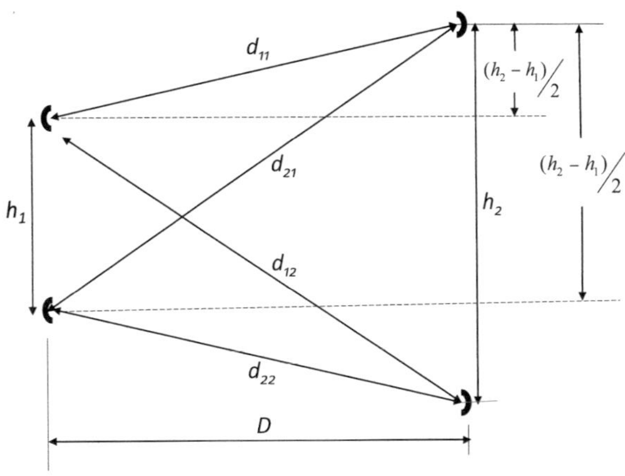

Fig. 6.6 2×2 LoS MIMO geometric data

$$d_{21} \simeq D + \left[\frac{h_2^2 + h_1^2 + 2h_1 h_2}{8D} \right] \tag{6.10}$$

Thus,

$$d_{21} - d_{11} = \frac{h_1 h_2}{2D} \tag{6.11}$$

Clearly, also

$$d_{12} - d_{22} = \frac{h_1 h_2}{2D} \tag{6.12}$$

For these path differences to be equal to $\lambda/4$, we get

$$\frac{h_1 h_2}{2D} = \frac{\lambda}{4} = \frac{C}{4f} \tag{6.13}$$

where C is the speed of light and equals 3×10^8 m/s, f is the frequency is c/s, and h_1 and h_2 are in meters. From Eq. (6.13), we can show that

$$h_1 h_2 = \frac{150 D_{km}}{f_{GHz}} \tag{6.14}$$

Where D_{km} is the path distance in kilometers and f_{GHz} is the frequency is GHz.

For the special case of equally separated antennas at both the transmit and receive side, we have

$$h_{opt} = \sqrt{\frac{150 D_{km}}{f_{GHz}}} \tag{6.15}$$

Figure 6.7 shows optimal antenna separation versus link length for a number of frequencies. As can be seen, optimum antenna separation when operating at 6 GHz is very large at typical link lengths for this frequency of about 20 to 40 km. Such separation and even non-optimum separation (see Sect. 6.3.4 below) may not be easily accommodated on the towers. As a result, the implementation of LoS MIMO at or near 6 GHz could prove challenging. At 80 GHz, on the other hand, where typical link length tends to be no more than about 5 km, the separation required is quite small and easily handled.

We note that with frequency division duplexed (FDD) links, where the go and return paths operate on different frequencies, antenna separation can never be fully optimal. Here "optimal separation" is normally based on the midfrequency between the operating frequencies.

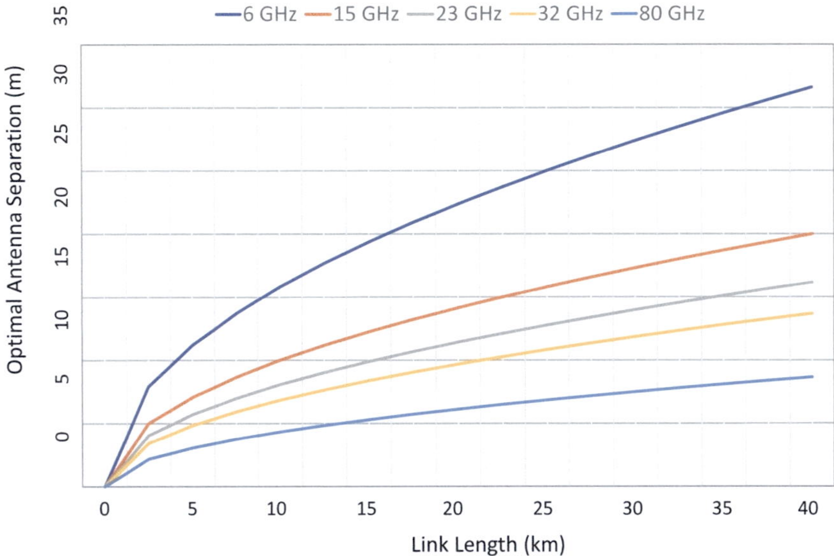

Fig. 6.7 Optimal antenna separation for a 2 × 2 Los MIMO system versus link length for a number of frequencies

6.3.4   *Non-optimal Antenna Separation*

In real-world installations, it is not always possible to mount antennas for optimum separation. This may be the case where the vertical separation called for is not available on the antenna tower/pole or where the horizontal separation required is not available on the rooftop supporting the antennas. When this is the case, sub-optimum separation must be employed. To get a sense of the effect of sub-optimal separation on performance we refer to Fig. 6.8 which depicts a LoS 2 × 2 MIMO system and where the antenna separation is equal on both ends. Here the separation is chosen to create a difference in the two paths from the transmitters to a receiver of $\lambda/8$. When this is the case then the antenna separation, $h_{sub\text{-}opt}$ say, can be shown to be equal to $h_{opt}/\sqrt{2} = .71d_{opt}$. Such a separation is a good trade-off between minimizing antenna spacing to meet practical limitations and, as we shall see, minimizing deterioration in SNR performance. In the figure, we see what action is required to eliminate interference. For the signal sent by Tx1, interference is eliminated by rotating a copy of the signal received by Rx2 by 135^0 and adding this to the signal received by Rx1. This action does indeed eliminate the interference from the signal sent by Tx2. It results in something else, however. The desired signal instead of adding coherently is added in vector form to give a signal that is of value $\sqrt{2}V$ instead of value $2\,V$ as was the case with optimum separation. This translates into a loss of relative signal power of 3 dB and hence no SNR/fade margin advantage relative to a SISO system, where the SISO transmitter power is the same as each

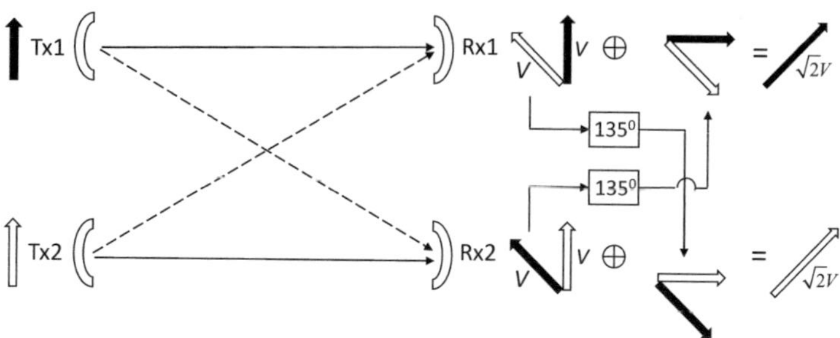

Fig. 6.8 Interference cancellation in a 2 × 2 LoS MIMO system with sub-optimal spacing

MIMO system transmitter power. Thus, we see that even with sub-optimal spacing it is possible to eliminate interference, albeit at the expense of SNR and hence fade margin.

6.3.5   LoS MIMO Equalization

There are a number of approaches to effect equalization of LoS MIMO systems, some analog, some digital, and operating at either RF or baseband. Following is a top-level review of some of these approaches.

The simplest approach conceptually is an analog one carried out in the time domain where a fixed analog network employing phase shifters or delay lines is used to create the signals necessary to cancel out inter-stream interference as shown, for example, in Fig. 6.5 and described in [8–10]. The performance of such equalization networks, however, is very sensitive to displacements of the antennas. This limitation can be addressed via some form of adaptive equalization employing variable phase shifters and variable gain amplifiers [9].

Another approach for a 2 × 2 system is one where equalization takes place in the frequency domain [11]. Shown in Fig. 6.9 is a simplified block diagram of the equalizer described in [11]. The two received signals first each undergo an FFT process that changes their representation from the time domain to the frequency domain. The FFT outputs then feed four frequency domain equalizers (FDEs) as shown. The FDEs are also fed by outputs from the communication path state estimator (CPSE). The two inputs to each FDE are then multiplied and the result outputted. FDE1 and FDE4 remove inter-symbol interference (ISI) caused mainly by frequency selective fading. The outputs of FDE 1 and FDE 2 are summed as are those of FDE 3 and FDE 4. The effect of this summing is to remove inter-stream interference. These summed outputs are then fed to IFFTs which convert their format from the frequency domain back to the time domain and the IFFT outputs feed conventional decoders. The CPSE is fed from the received signals and processes training sequences sent from each transmitter in order to generate the information sent to each FDE.

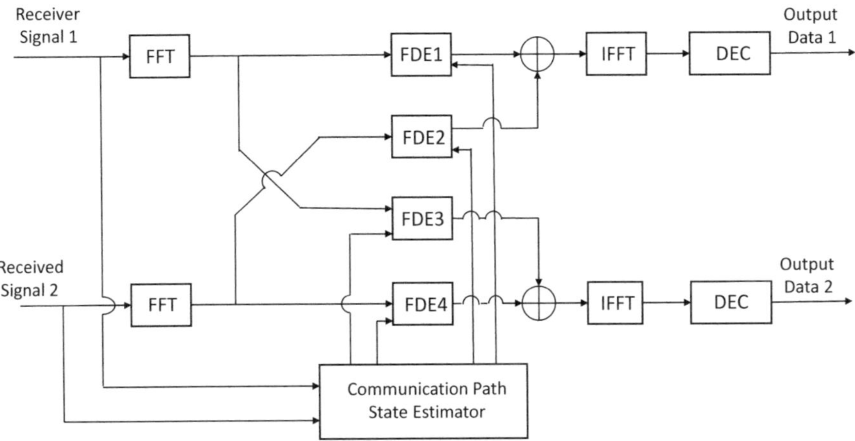

Fig. 6.9 Frequency domain equalization

The more common approach to LoS MIMO equalization is to use DSP-based equalizers. Such equalizers, like those described immediately above, work not only to cancel inter-stream interference but also to remove ISI and are referred to as space-time equalizers (STEs). These equalizers, just like those for SISO systems that address only ISI, must not only be able to initialize FIR filter taps but also be able to update these taps iteratively as the channel conditions change over time. As was discussed with XPIC, initialization can be achieved via training sequences, but this is not the preferred approach with LoS systems. Rather, blind equalization [3] is preferred. Shown in Fig. 6.10 a simplified block diagram of such a system. Filter taps are updated based on algorithms such as the Modified Constant-Modulus Algorithm (MCMA) [4] and the simplified constant-modulus algorithm (SCMA) [5] which were discussed in Sect. 5.9.2.1. Following initialization of the filter taps the Decision-Directed mode is commonly used where algorithms such as the least mean square (LMS) [2] and the Recursive Least Square (RLS) [2] are used which were also discussed in Sect. 5.9.2.1. Shown in Fig. 6.11 a simplified block diagram of such a system.

6.3.6   *Increasing Channel Capacity Via the Simultaneous Use of CCDP/XPIC and LoS MIMO*

As was shown above, 2×2 LoS MIMO doubles channel capacity by transmitting two independent signals over the same spectrum allocation with the same polarization. It is possible, however, to employ co-channel dual polarization transmission along with 2×2 LoS MIMO where each of the two antennas at each end is a dual-polarized one. Such a configuration results in the transmission of four independent

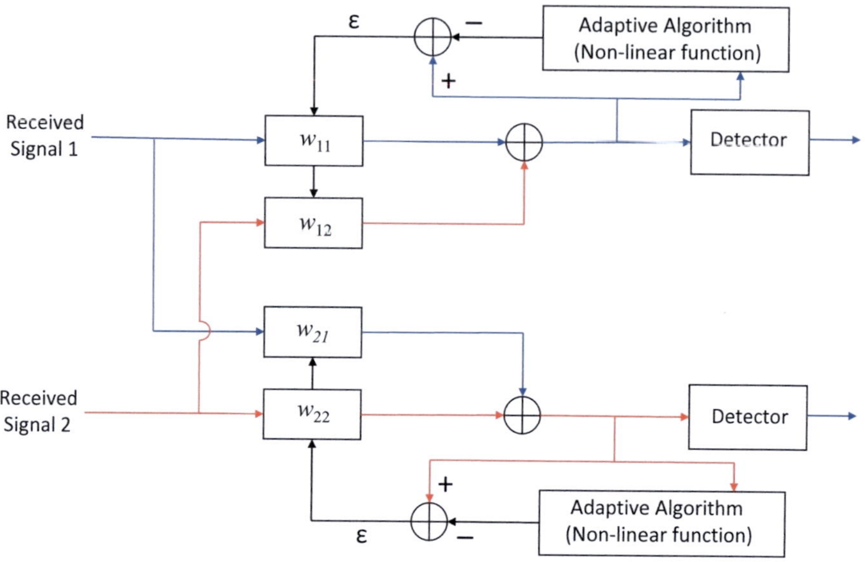

Fig. 6.10 LoS MIMO equalizer under the blind equalization mode

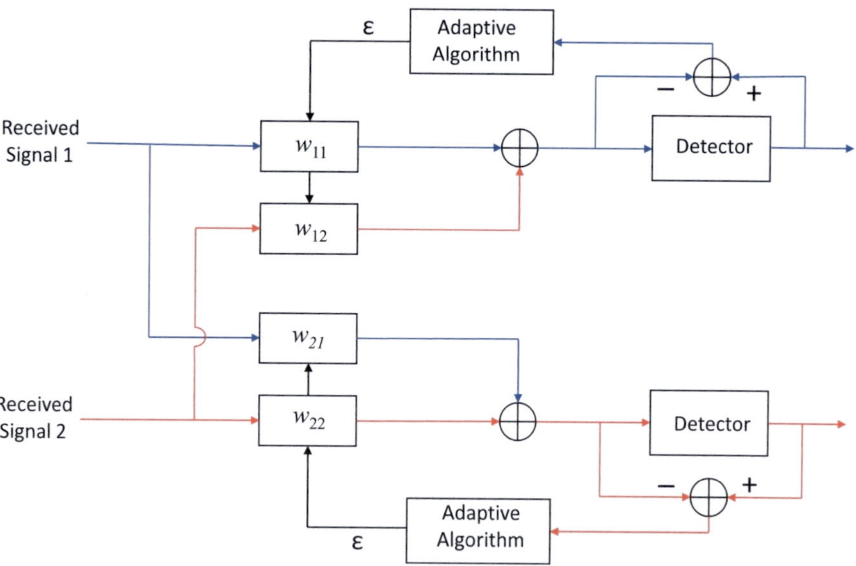

Fig. 6.11 LoS MIMO equalizer under the decision-directed mode

signals leading to a quadrupling of capacity relative to a single polarization SISO system and is often referred to as a 4×4 LoS MIMO one as it is based on four independent transmitters and four independent receivers.

At the receiving end of such a configuration, not only must the spatially received MIMO interfering signals be eliminated, but also the cross-polarized signals must be cancelled via XPIC. The adaptive *space-time equalizer* (STE) required here becomes quite complex. A block diagram of an STE for such an arrangement is given in [12] and shown in Fig. 6.12. "The analog signals from the antennas are sampled and filtered before entering the STE which is comprised of 16 FIR filters. Filters w_{11}, w_{22}, w_{33}, and w_{44} remove the ISI, while filters w_{13}, w_{31}, w_{24}, and w_{42} remove the co-polarization interference for each polarization state. The remaining filters combat the cross-polarization interference and are referred to as cross-polarization interference cancellers (XPICs)" [12].

Because it is easier and more cost-effective to increase capacity via co-channel dual polarization (CCDP) as opposed to LoS MIMO, CCDP is often used ahead of LoS MIMO. When this is done, then should additional capacity be required, LoS MIMO can be then added.

Though the discussion above has been largely centered on the 2×2 antenna per end, structure, more antennas per end, and hence greater capacity increase, is possible. For the 4×4 antenna per end structure, the antennas at each end can be arranged either in an equally spaced colinear fashion or on the four corners of a square.

6.4   Orbital Angular Momentum Multiplexing

6.4.1   Introduction

Orbital angular momentum (OAM) of electromagnetic waves was discovered in the 1990s and found initial application in the field of optical transmission. It has theoretically an infinite number of transmission modes that are mutually orthogonal to each other. Utilizing this feature can allow a system to transmit multiple waves that are coaxially propagating and spatially overlapping, each wave conveying an independent data stream. Just prior to the 2010s researchers started to study the application of OAM multiplexing to radio transmission and throughout the 2010s experimental work was carried out by many that demonstrated the feasibility of this technological approach. The application, however, is not without its limitations. In particular, it is limited in its ability to convey data over distances measured in kilometers when using antenna sizes similar to those used for traditional transmission. For mobile network wireless transport, link distances from a low of about 100 m to a high of 10s of kilometers are typically required. Research and development in the OAM arena for mobile wireless transport is directed towards viable links operating at the lower end of these distances. It is not clear at this time if OAM multiplexing will in fact find a viable role in wireless transport alongside proven technologies such as co-channel dual polarization transmission and LoS MIMO. However, given its potential to aid in capacity increase, a cursory review of its structure, characteristics, and a possible method of physical realization seems to be in order.

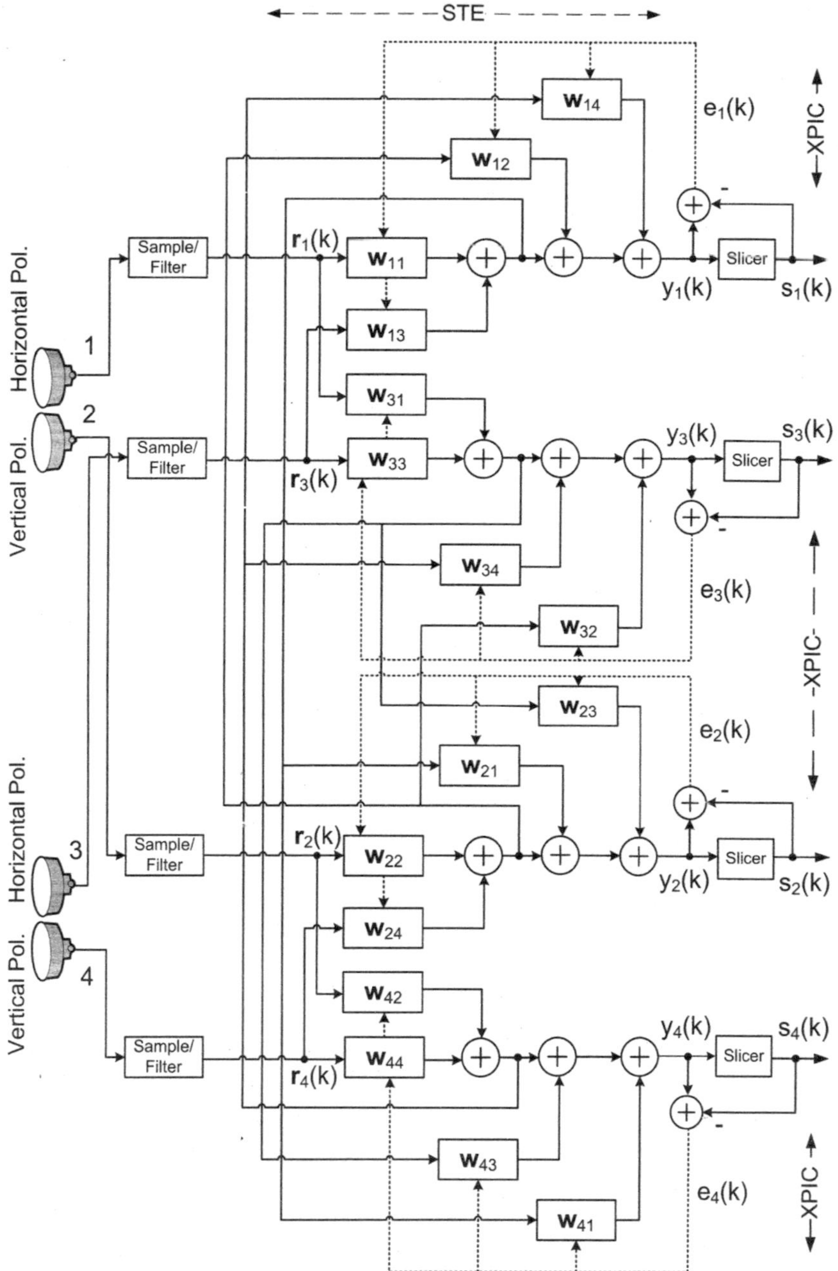

Fig. 6.12 Block diagram of a 4 × 4 spatially separated, dual, polarized Los MIMO receiver. (From [12], with the permission of the IEEE)

6.4.2   OAM Structure and Characteristics

Traditional wireless communications, from the time of Marconi, are based on the application of the plane-electromagnetic (PE) wave which possesses not only linear momentum but spin angular momentum (SAM), the latter being associated with the state of polarization of the beam. Electromagnetic waves can possess, however, not only linear momentum and SAM but also orbital angular momentum (OAM) which impacts the spatial phase profile of the beam which now presents itself as a wavefront with a helical phase. Beams in the OAM mode have a $e^{il\phi}$ phase factor, where i equals $\sqrt{-1}$, l is an integer indicating the OAM mode number and can be of any positive or negative value, and ϕ is the azimuthal angle. The wavefront phase rotates around the direction of beam propagation with the phase changing $2\pi l$ after a full turn. OAM in theory can have infinite independent modes (0, +/− 1, +/− 2, ...). These modes with different values of l are mutually orthogonal, thus signals with different modes can be multiplexed at the transmit end, transmitted over the same path utilizing the same transmit and receive antennas, and demultiplexed at the receiving end with low inter channel interference. The wavefront of the OAM signal is different for all modes. These differences can be seen in Fig. 6.13 [13] which shows the wavefronts of OAM waves with modes 0, +1, +2, and + 3.

The power density distribution of an OAM signal has a ring shape, the radius of which increases as the mode order increases, as the distance from the transmit antenna increases, and as the transmission frequency decreases. The larger the beam divergence, the larger the required receive antenna to collect the transmitted power or, should the antenna not be large enough, the lower the SNR at the receiver. The relationships governing ring radius are clearly seen in Fig. 6.14 [14] and are the reason why, as indicated above, OAM transmission is limited in its ability to convey data over distances measured in the kilometers when using antenna sizes similar to those used for traditional transmission. It should be noted that in mode 0 no orbital angular momentum is imparted. Thus, strictly speaking, this mode is not an OAM mode, but rather represents standard PE wave transmission and thus there is no OAM generated beam divergence.

The increase in beam divergence with OAM mode order puts a practical limit on the number of modes that can be transmitted simultaneously. This thus begs the question as to whether practical OAM multiplexed transmission can provide greater

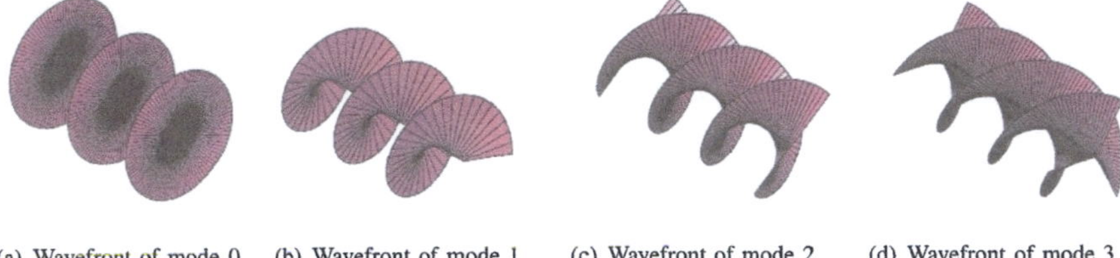

(a) Wavefront of mode 0.   (b) Wavefront of mode 1.   (c) Wavefront of mode 2.   (d) Wavefront of mode 3

Fig. 6.13 Wavefront for OAM waves with different mode. (From [13], with the permission of the IEEE)

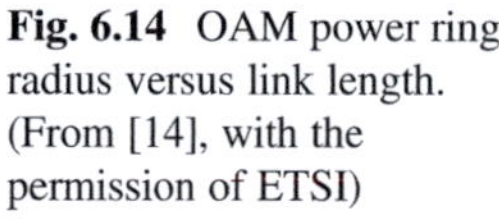

Fig. 6.14 OAM power ring radius versus link length. (From [14], with the permission of ETSI)

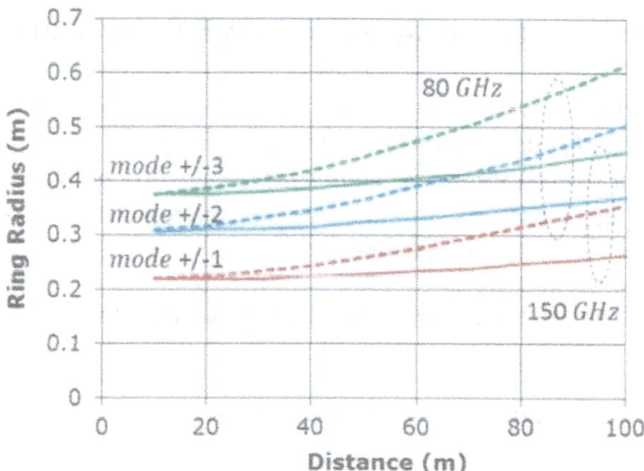

capacity than LoS MIMO. Regardless of the answer, it turns out that MIMO multiplexing can be used jointly with OAM multiplexing [15, 16], and such an arrangement holds the promise of greatly increases spectral efficiency. To add to the potential complexity, it has been shown that waves can exhibit OAM and the different polarization states simultaneously. Thus, the successful combination of OAM multiplexing and polarization multiplexing would also lead to very high spectral efficiency. The ultimate combination would obviously be one that encompassed OAM, dual-polarization, and LoS MIMO multiplexing.

Because of OAM ring radius dependence on link length and transmission frequency, the likely application of OAM in 5G wireless transport, if proven feasible, will for links on the order of 100 m and operating in the high millimeter-wave bands, specifically the E-band (71–86 GHz), and when feasible the W-band (92–114 GHz), and the D-band (141–175 GHz).

6.4.3 OAM Mode Generation and Multiplexing/Demultiplexing

The key technologies required to effect OAM-based wireless transmission are those associated with mode generation and data signal multiplexing/demultiplexing. Much work has been done in this area since the early 2010s. One approach of such technologies is to first generate single OAM modes then multiplex them. The second is to generate all the desired modes simultaneously and coaxially aligned on a *uniform circular array* (UCA) that's driven by a *Butler matrix* [17]. This second approach is considered to be the more suitable and is where much research effort is directed. We will thus examine it at a high level.

A UCA is an array of N sub-antennas located on a circle equally spaced from one another. The kth sub-antenna element is driven by a signal with an $k \cdot \theta_l$ phase delay, where

$$\theta_l = (2\pi l)/N \tag{6.16}$$

Here, l is the OAM mode, and N must be such that $|l|$ is an integer smaller than $N/2$. With these conditions met, the wavefront will contain the desired $e^{il\phi}$ phase factor. By using a UCA, each sub-antenna element can be driven by a multiplexed signal. Thus, a single UCA can create a transmission beam with multiple OAM modes if driven by the appropriate synthesized signals. Such signals are created by a Butler matrix.

An N x N Butler matrix is used to drive a UCA with N sub-antennas. For each input signal to the matrix, there are N output signals with phase delays proportional to the sub-antenna number, the phase delay increment being defined by the input port number. These N output signals feed the N antenna ports. With different signals present simultaneously, each feeding a different input port, the matrix outputs the required composite signals to result in the UCA generating the desired OAM beams.

To get a sense of how this UCA/Butler matrix arrangement works consider a UCA with 8 sub-antennas and hence an 8×8 associated Butler matrix. For a Mode 1 OAM beam, the phase delay increment, as determined by Eq. 6.16, is 45^0. Thus, for a signal at the Butler matrix input port designated as that supporting Mode 1, the matrix outputs the 8 signals of relative phases as shown in Fig. 6.15a. Also shown in Fig. 6.15a are the relative phases of a Mode 2 transmission. Because here N equals 8, $|l|$ must be an integer smaller than 4. Thus, only modes 0, $+/-$ 1, $+/-$ 2, and $+/-$ 3 are supported. Should all supported modes be utilized, then the matrix would have 7 inputs and thus one input would not be utilized.

At the receiving end, the separation of beams carrying OAM modes can be accomplished in a fashion similar to that used for generation. This is accomplished by using a UCA with the same number of sub-antennas as at the transmit end but with the sub-antennas connected with phase shifters that result in phase rotation in the opposite direction to that imparted at the transmit end. This is shown in Fig. 6.15b for Mode 1 and 2 beams.

6.5 Band and Carrier Aggregation

When using all the standard techniques to boost capacity over a given channel still does not provide the desired capacity at the desired reliability, wireless transport providers often turn to multi-channel solutions. These solutions can be single-band or multiband and are referred to as band and carrier aggregation [18]. As per [18], "band and carrier aggregation n (BCA) is a concept enabling an efficient use of the spectrum through a smart aggregation, over a single physical link, of multiple frequency channels (in the same or different frequency bands)."

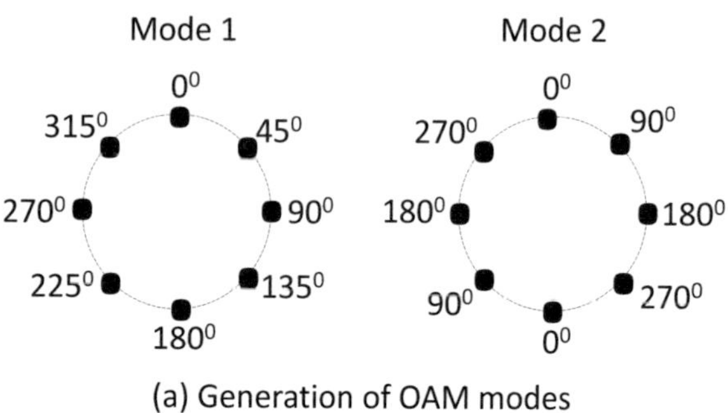

Fig. 6.15 Generation and separation of OAM modes with a UCA

(a) Generation of OAM modes

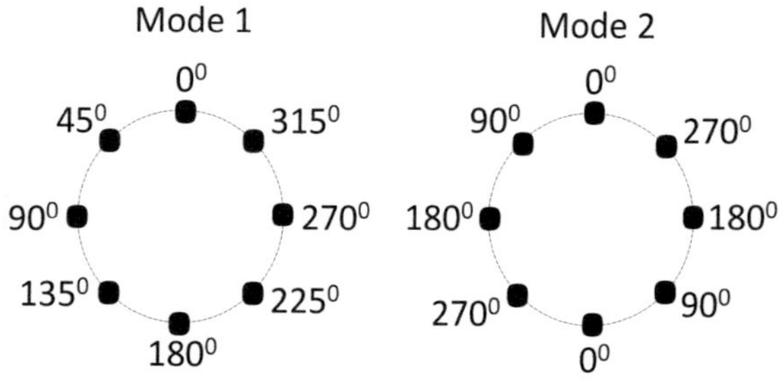

(b) Separation of OAM modes

In the single-band solution, multiple channels in a single frequency band are used simultaneously, with the user data divided between them and the net impact is an increase in capacity. In the multiband solution, one or more channels in a given frequency band are combined with one or more channels in multiple alternate frequency bands, though typically the combination involves only two bands. The multiband solution is particularly powerful because it not only allows an increase in capacity, but if properly planned, it allows an increase in availability and, if necessary, an increase in hop length.

The multiband solution, an example of which is illustrated in Fig. 6.16, works best when ultra-high capacity is made available from a channel or channels in the high millimeter-wave bands, the E-band being particularly favored here, and high availability is made available from a lower band, typically 15, 18, or 23 GHz. Availability is impacted negatively by rain, and the higher the frequency, the greater the impact. It is for this reason that a large difference in frequency between the two bands is important. With this solution, the channels, particularly the higher frequency one, typically employ adaptive modulation and possibly adaptive coding to maintain maximum possible capacity as propagation conditions change. User data is adaptively re-routed among the different channels based on the instantaneously

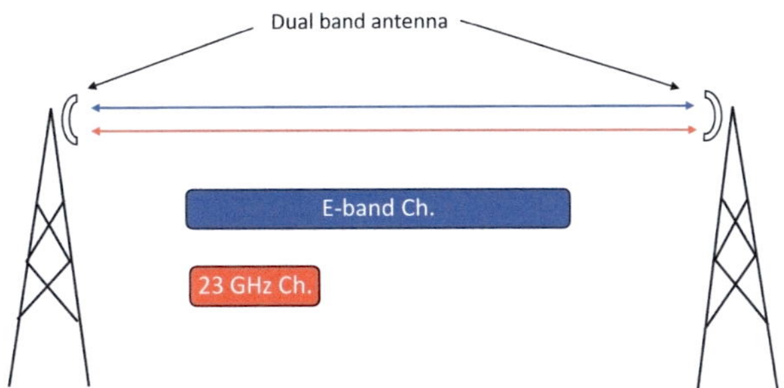

Fig. 6.16 Band and carrier aggregation example

available capacity of the channels and, as capacity changes, a hitless redistribution of data takes place automatically taking into account traffic priority.

To get a sense as to how the multiband solution operates, let's consider the hypothetical situation where the higher frequency band is the E-band and the lower band the 23 GHz one. The E-band channel bandwidth is 500 MHz, uses 256-QAM, and the maximum link capacity is 3.4 Gb/s. The 18 GHz channel bandwidth is 56 MHz, uses 4096-QAM, and maximum link capacity is 600 Mb/s. Thus, when combined and each channel transmits data at its maximum rate, the maximum combined data rate is 4 Gb/s. The key design consideration here is the achievement of the desired availability. Let us assume that the hop length is such that the E-band link has an availability of 99.9% and the 23 GHz link an availability of 99.999%. This translated to being able to maintain a capacity of at least 600 Mb/s with 5 min of downtime per year and a capacity of 4 Gb/s for all but 8.8 h per year. Further, high priority traffic is likely prioritized to have access to the 600 Mb/s capacity during those 8.8 h. What if we needed to communicate over a path length somewhat longer than the one implied above. If nothing but the path length changed then the availability on both bands would go down. To assure a continued minimum availability of 99.999% for a capacity of 600 MHz we could reduce the lower frequency to 15 GHz say, or we could increase the size of the lower frequency antennas. Figure 6.17, which is from [19], shows achievable distances with high-capacity wireless links for different climates and levels of availability. The mild climate is for where a rain rate of 30 mm per hour is exceeded for 0.01% of the year and the severe climate is for where a rain rate of 90 mm per hour is exceeded for 0.01% of the year. The availability targets in the 6–42 GHz range are set to half the maximum link capacity which corresponds to transmission at 64-QAM for links capable of operating at 4096-QAM. In the 60–80 GHz range, the availability targets are also set to half the maximum link capacity which corresponds to transmission at 16-QAM for links capable of operating at 256-QAM. Obviously, full link capacity has lower availability than that shown but is nonetheless available for most of the year. The figure shows two multiband potential arrangements, one for links with lengths of up to about 5 km and the other for links with lengths of up to about 15 km.

An important innovation that assists in multiband transmission is multiband antennas which facilitate multiband channels on a single dish, thus minimizing the required tower space, wind loading, and installation and rental cost. Use of such

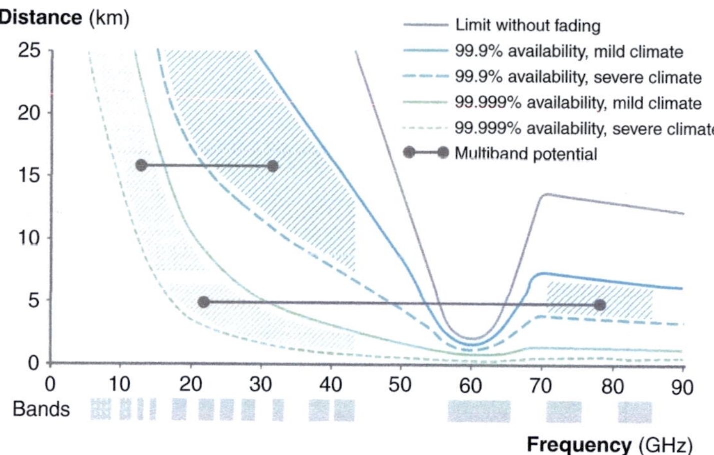

Fig. 6.17 Achievable distances with high-capacity wireless links. (From [19], with the permission of Ericsson)

antennas, however, when band separation is high, is not without trade-offs. As antenna gain increases the beamwidth decreases. As a result, gain should not exceed the low 50s in dBs as a practical rule, since beyond this antenna alignment becomes very challenging. The multiband antenna size is thus limited by the highest frequency used since a gain close to 50 dB will be achieved at a lower antenna diameter for the highest frequency relative to the lowest one (see Fig. 3.4). For an 80 GHz/ 23 GHz multiband antenna, for example, this implies a maximum antenna diameter of 0.6 m with a nominal gain at 80 GHz of 51 dB. At 23 GHz this diameter would result in a gain of only about 40 dB thus putting a limit on the fade margin and hence availability achievable at this frequency. A further trade-off is that antenna gain is impacted negatively, with the actual gain for an E-band/38 GHz antenna being 2 dB lower compared to traditional single-band antennas [18]. In Chap. 7, path analyses are presented that shown the availability of an 80/32 GHz aggregated link.

6.6 Summary

In this chapter, techniques employed in point-to-point links in order to enhance capacity were described which, when used along with modulation-based capacity improvement, can lead to extremely high-capacity capability. Among the techniques described were (a) co-channel dual polarization (CCDP) transmission accompanied by cross-polarization interference cancellation (XPIC) in the receiver, (b) line-of-sight multiple-input multiple-output (LoS MIMO) transmission, and (c) band and carrier aggregation (BCA). These techniques are fully developed, being widely employed, and can be applied individually. However, it is when they are applied collectively that a truly great increase in capacity can be achieved. Orbital angular momentum (OAM) multiplexing was also described. It, unlike the above, is still very much in the research and development phase, and the jury is out as to whether it will find a significant place in wireless transport systems.

References

1. Cui L et al (2012) Research on cross-polarized interference canceller with blind adaptive algorithm, international conference on computational problem solving. Leshan, China
2. Proakis JG (2008) Digital communications, 5th edn. McGraw Hill, New York
3. Garth L et al (1998) An introduction to blind equalization, TD-7 of ETSI/STS TM6. Spain, Madrid
4. Oh KN, Chin YO (1995) Modified constant modulus algorithm: blind equalization and carrier phase recovery algorithm. Proc IEEE Int Conf Commun 1:498–502
5. Lkhlef A, Guennec DL (2007) A simplified constant Modulus algorithm for blind recovery of MIMO QAM and PSK signals: a criterion with convergence analysis. Eurasip J Wirel Commun Netw 2007, Article ID 90401, pp 1–13
6. CEPT Electronic Communications Committee (2017) ECC report 258: guidelines on how to plan for point-to-point fixed service links. Copenhagen, Denmark
7. Ingason T, Haonan L (2009) Line-of-sight MIMO for microwave links: adaptive dual polarized and spatially separated systems. Chalmers University of Technology, Goteborg
8. Sheldon S et al (2008) A 60 GHz line-of-sight 2×2 MIMO link operating at 1.2 Gbps. In: IEEE antennas and propagation society international symposium, San Diego, California
9. Song X, et al (2016) Analog and successive channel equalization in strong line-of-sight MIMO communication. In: IEEE International Conference on Communications, Kuala Lumpur, Malaysia
10. Song X, et al (2015) Strong LOS MIMO for short range MmWave communication. In: IEEE International Conference on Ubiquitous Wireless Broadband, Montreal, Canada
11. Kamiya N (2019) LOS-MIMO demodulation apparatus, communication apparatus, LOS-MIMO transmission system. In: LOS-MIMO demodulation method and program, United States Patent Application Publication Pub. No. US 2019/0020384 A1
12. Ingason T et al (2010) Impact of frequency selective channels on a line-of-sight MIMO microwave radio link. In: IEEE 71st Vehicular Technology Conference, Taipei, Taiwan
13. Cheng W et al (2019) Orbital angular momentum for wireless communications. IEEE Wirel Commun 26(1)
14. ETSI (2018) Millimeter Wave Transmission (mWT); Analysis of Spectrum, License Schemes and Network Scenarios in the D-band, ETSI GR mWT 008 v1.1.1, ETSI, Sophia Antipolis, France
15. Ren Y et al (2017) Line-of-sight millimeter-wave communications using orbital angular momentum multiplexing combined with conventional spatial multiplexing. IEEE Trans Wirel Commun 16(5):3151–3161
16. Cheng W et al (2017) Orbital-angular momentum embedded massive MIMO: achieving multiplicative spectrum-efficiency for mmWave communications. IEEE Access 16:2732–2745
17. Lee W et al (2017) Microwave orbital angular momentum mode generation and multiplexing using a waveguide Butler matrix. ETRI J 39(3): pp. 336–344
18. ETSI (2017) Frequency bands and carrier aggregation systems; Band and Carrier Aggregation, ETSI GR mWT 015 v1.1.1, ETSI, Sophia Antipolis, France
19. Edstam J (2016) Microwave Backhaul gets a Boost with Multiband, Ericsson Technology. Review, January 2016

Chapter 7
Transceiver Architecture, Link Capacity, and Example Specifications

7.1 Introduction

The architecture of wireless transport transmitters and receivers is not rigid but can take many forms and many physical manifestations. How the designer chose to structure it is influenced by many factors including transmission data rate, transmission frequency, operating environment, etc. In this chapter, we will explore the basic transceiver architecture and structural options thereof. We then examine the capacity capability of modern wireless transport links. Finally, to demonstrate the impact on the performance of the technologies presented in previous chapters, we review the high-level architecture and some key specifications of two current wireless transport systems, one operating in the traditional band and one in the high-millimeter-wave E-band.

7.2 Basic Transceiver Architecture and Structural Options

Shown in Fig. 7.1a is a highly simplified representation of a wireless transport transceiver that employs *direct conversion* (*homodyne conversion*). It comprises a *baseband processor* (BBP), an RF front end, an antenna coupler, and an antenna. In this interpretation, the BBP is responsible for all processing in the transmit direction up to the creation of the I and Q data streams necessary for *quadrature amplitude modulation* (QAM) and all processing in the receive direction after the creation of the QAM demodulated I and Q data streams. The *RF front end* is responsible on the transmit side for creating the RF QAM signal from the I and Q data streams via direct conversion, summing, and amplification. On the receive side, it is responsible for the low-noise amplification of the received QAM signal and coherent demodulation thereof into its I and Q components. The antenna coupler allows the transmitter and

D. H. Morais, *5G and Beyond Wireless Transport Technologies*,
https://doi.org/10.1007/978-3-030-74080-1_7

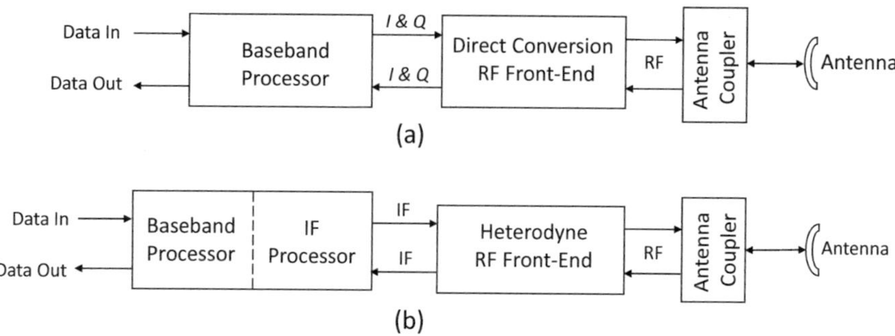

Fig. 7.1 Wireless Transport transceiver block diagram (**a**) Direct Conversion (Homodyne Conversion) (**b**) Heterodyne Conversion

receiver to access the antenna via a single port. The antenna is responsible for radiating the outgoing signal while simultaneously capturing the incoming signal.

In wireless transport transceivers, direct conversion is the go-to approach for high-bandwidth (100 s of MHz), high-millimeter-wave transmissions such as in the V-band and E-band. This is because in these bands very high-bandwidth channels are available (up to 2 GHz in the E-band) resulting in very high data rate I and Q streams that would in turn necessitate excessively wide IF bandwidths that would complicate the design of proper functioning circuitry coupled with very high IF frequencies. Further, because linear transmission of these very high I and Q baseband data streams over any distance is difficult, the BBP and the RF front end are usually collocated. Finally, as transmission loss over waveguide at these frequencies is high, the RF front end is normally attached directly or via an extremely short connection to the antenna via the antenna coupler. The net result is that the entire transceiver components coalesce into one integrated outdoor mounted unit (Full Outdoor Mount). As the data in and out of the transceiver is Ethernet-based, the connections between the transceiver and the external interface, located near the base of the tower, is provided via Ethernet cable which also provides power to the transceiver (See Fig. 7.2a).

Shown in Fig. 7.1b is a highly simplified representation of a wireless transport *heterodyne* conversion transceiver and hence where QAM IF signals are employed. Here the baseband processing is identical to that for the direct conversion option. However, now the baseband processor interfaces with an integrated IF processor that is responsible on the transmit side for creating an IF QAM signal from the I and Q data streams and passing this signal onto the RF front end and on the receive side for the coherent demodulation of the received IF signal from the RF front end into its I and Q components and passing these signals onto the BBP. The RF front end here is responsible on the transmit side for *upconversion* to the desired RF frequency following by amplification and on the receive side for low-noise amplification followed by *downconversion* to the IF frequency. The function of the antenna couplers here is the same as in the direct conversion option above.

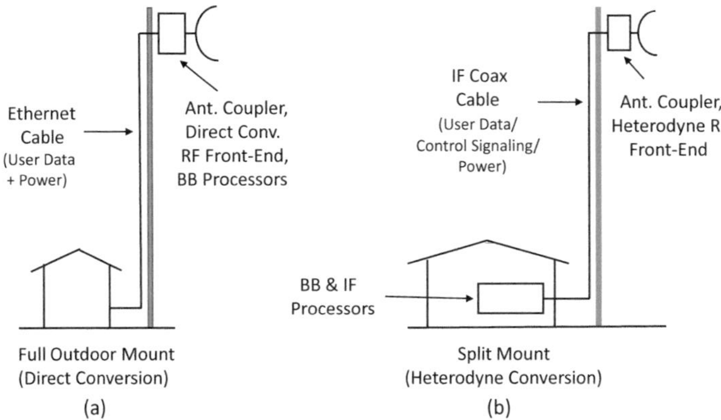

Fig. 7.2 (**a**) Full outdoor mount configuration, (**b**) Split mount configuration

In wireless transport transceivers, the use of an architecture encompassing an intermediate frequency is the common approach for systems operating in the traditional bands and the high-millimeter-wave bands where the channel bandwidth is normally not much more than 112 MHz for example. Here many physical locations of the modules are possible. One possibility is to have all modules outside and collocated. Another possibility, as shown in Fig. 7.2b, is to have the RF front end outside adjacent to the antenna and the baseband and IF processor located indoors (split mount). Finally, both the baseband and IF processor as well as the RF front end can be located indoors and connected to the antenna via a waveguide run. This last option is usually used when the operating frequency is towards the low end of the traditional band, at 6 GHz for example, and where the transmit output power is several watts, thus requiring high input power.

Shown in Fig. 7.3 are a number of Ericsson wireless transport terminals and two associated parabolic antennas mounted on a pole which is in turn rooftop mounted. Specifically, the figure shows (from top to bottom) ML6363 with integrated antenna, ML6352 (E-band) with integrated antenna, ML6366 (all-outdoor radio), and an associated parabolic antenna, a dual-band antenna with an ML6363 (lower frequency) and an ML6352 (E-band), and an Ericsson AIR radio (a 5G NR antenna-integrated radio) at the bottom.

7.2.1   *The Baseband Processor*

Shown in Fig. 7.4 is a simplified block diagram of the Baseband Processor depicted in Fig. 7.1a, b. The input data on the transmit side is of Ethernet format which here for simplicity we have assumed has already undergone multiplexing, if necessary, and header and payload compression. Similarly, the output data on the receive side is assumed to be header and payload compressed. The Ethernet traffic interface with

Fig. 7.3 Wireless transport terminals and associated parabolic antennas. (With the permission of Ericsson)

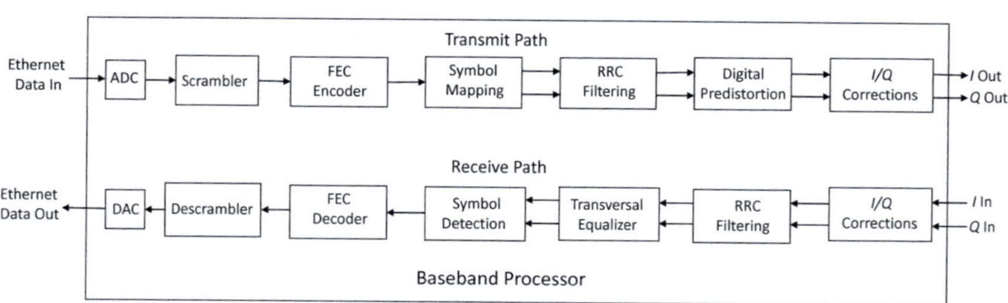

Fig. 7.4 Baseband processor

external traffic (before header and payload compression and multiplexing/after header and payload decompression and demultiplexing) can be either electrical or optical (see Sect. 2.2.5). All such modern processors are DSP-based, and thus all processing takes place digitally. As a result, the input to the processor on the transmit side, even though it is a digital stream of data, is further digitized by an *analog to digital converter* (ADC) to a much higher data rate where the digital processing takes place. The input to the processor on the receive side is digitized, thus just prior to the output, a *digital to analog converter* (DAC) is used to restore the information to its original data format. Other than the ADC and the DAC, all the functions shown in Fig. 7.4 have been covered in preceding chapters. Specifically, these functions have been covered in the sections indicated below:

Scrambiling/descrambling: Sect. 4.5.1.
FEC encoding/decoding: Sect. 5.2.
Symbol mapping/demapping: Sect. 4.3.

Root raised cosine (RRC) filtering: Sect. 4.2.
Digital predistortion: Sect. 5.6.
I/Q corrections: Sect. 5.8.

7.2.2   The IF Processor

Shown in Fig. 7.5 is a simplified block diagram of the IF Processor depicted in
Fig. 7.1b. Processing in this module takes place in the analog domain. As the outputs
and inputs to the BBP are all digital, DACs [1] are required to convert the inputs
from the BBP to analog and ADCs [1] are required to convert the outputs to the BBP
to digital. On the transmit side, an IF local oscillator drives two mixers, one fed by
the I data stream, the other by the Q data stream, to create via summing a QAM
signal at the IF frequency. On the receive side, the incoming IF signal is split and fed
to two mixers each also driven by an IF local oscillator, thus creating I and Q data
streams. The IF diplexer connects the IF processor to the heterodyne RF front end
depicted in Fig. 7.1b via a single coaxial cable. It is able to do this because a different
IF frequency is used in the Go direction than in the Return direction. This cable also
normally provides power to the RF front end as well as conveying some control
signaling.

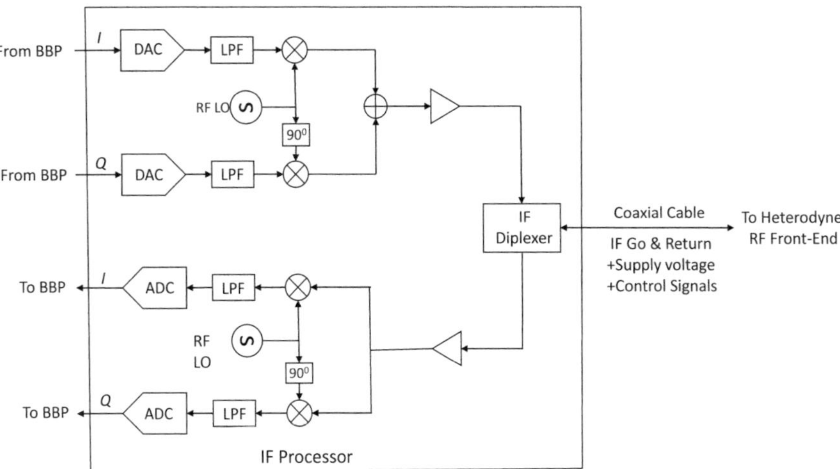

Fig. 7.5 IF processor

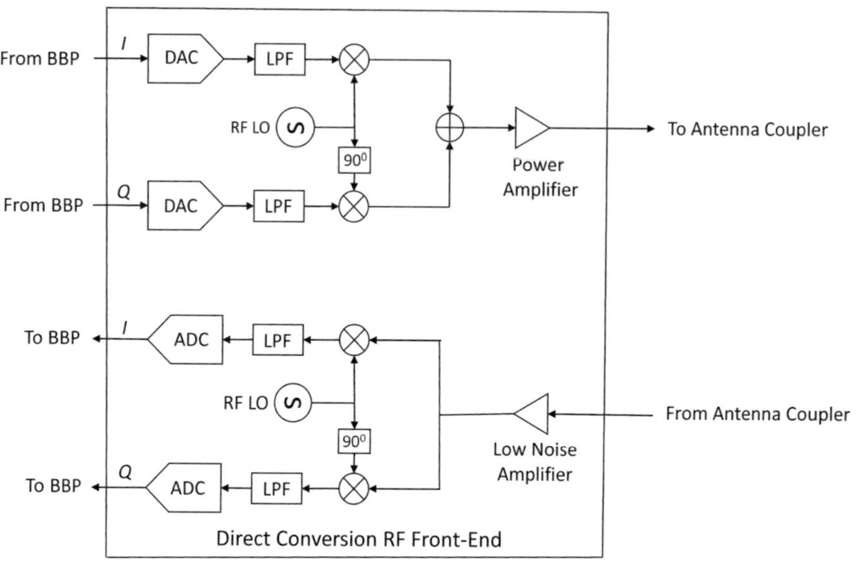

Fig. 7.6 Direct conversion RF front end

7.2.3 *The Direct Conversion RF Front End*

Shown in Fig. 7.6 is a simplified block diagram of the direct conversion RF front end depicted in Fig. 7.1a. We note that the coherent modulation and demodulation are very similar to those in the IF processor, the difference being that the local oscillators operate at RF at the actual transmission frequencies. As with the IF processor, DACs and ADCs [1] are required to effect the necessary interface with the BBP.

7.2.4 *The Heterodyne RF Front End*

Shown in Fig. 7.7 is a simplified block diagram of the heterodyne RF front end depicted in Fig. 7.1b. The front-end interfaces with the IF processor on one side and with the antenna coupler on the other. On the transmit side, it upconverts the received IF signal to RF and then amplifies it, whereas on the receive side, it amplifies the received RF signal and then downconverts it. Up- and downconversion were covered in Sect. 4.4.1.

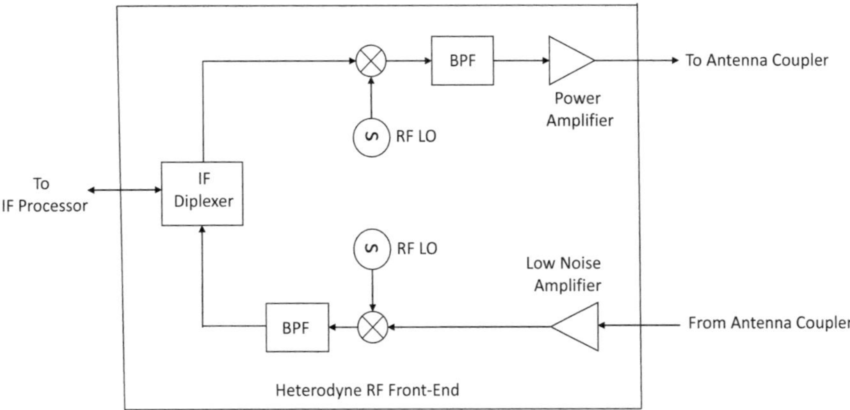

Fig. 7.7 Heterodyne RF front end

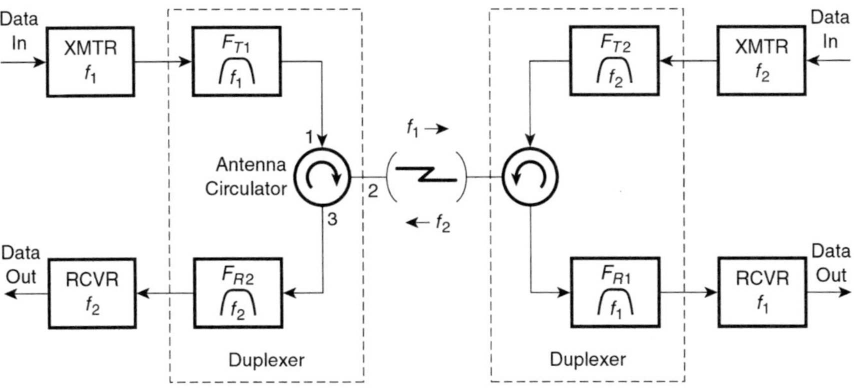

Fig. 7.8 Single-channel antenna duplexing

7.2.5 *Antenna Coupling*

For a single-channel (two-way) terminal, antenna coupling takes the form of an *antenna duplexer*, as shown in Fig. 7.8. A key component of the duplexer shown is the antenna circulator. A basic circulator is a three-port device, constructed from ferrite material, with behavior such that an input signal to any port circulates unidirectionally and exits at the next port on the unidirectional path. Thus, in the figure, the transmit signal of frequency f_1 that enters port 1 of the antenna circulator exits port 2 and proceeds via transmission line to the antenna. Likewise, the signal received by the antenna of frequency f_2 enters port 2, exits port 3, and proceeds to the receiver. Because the transmit signal can never be totally absorbed by the antenna system, a small fraction of it returns to port 2 and ends up at the receiver input. The duplexer filter, F_{T1}, in the transmit leg, limits the level of noise and spurious emission that falls within the receiver passband. It is designed to do this to a degree

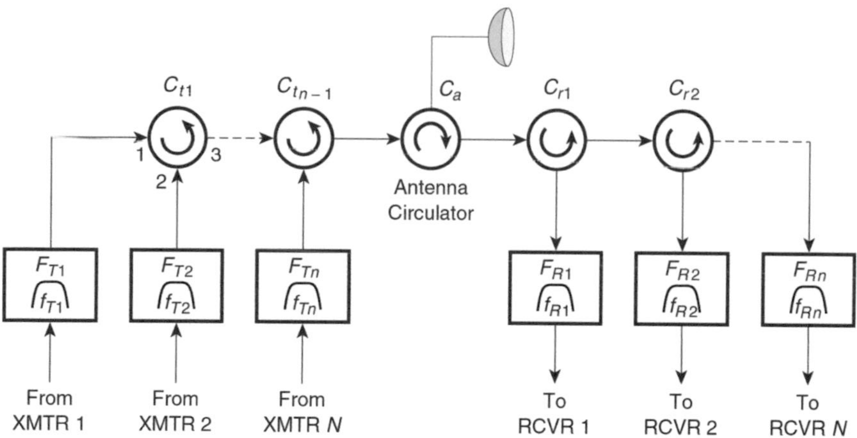

Fig. 7.9 *N*-channel antenna coupler

that any such unwanted input that appears at the receiver front end is at a level low enough as to not degrade receiver BER performance. The filter F_{R2} in the receive leg of the duplexer helps ensure that the level of the transmitted signal reaching the receiver front end does not overload it, resulting in non-linear behavior that degrades the BER performance. The loss experienced by a transmit or receive signal through a duplexer varies depending on the design but typically runs between about 1 and 2 dB.

For terminals supporting more than one channel on the same polarization, the antenna coupler is an expanded version of the duplexer and is structured as shown in Fig. 7.9. By the use of additional circulators and filters, branching networks are created on each side of the antenna circulator. The signal from transmitter 1, S_{T1} say, passes through filter F_{T1}, into port 1 of branching circulator C_{t1}, out of its port 2 and to the input of filter F_{t2}. The input of F_{t2} is reflective to signals outside its passband. As a result, S_{T1} is reflected back to port 2 of C_{t1}, reenters it, exits port 3 and continues on in a similar fashion until it exits the antenna circulator, C_a. On the receive side, receive signals behave similarly to transmit signals on the transmit side, being reflected off filters whose passbands are removed from the signals occupied bands.

Often single-channel systems are operated in an equipment protection mode. A common version of such protection is shown in Fig. 7.10 and is referred to as *monitored hot standby protection*. In such a scheme, two fully operational transmitters and receivers are employed at each terminal. On the transmit side both transmitters are modulated with the input data, but only one, transmitter *A* in the figure, is connected to the antenna duplexer, this transmitter being referred to as the working transmitter. The other transmitter, which is referred to as the standby transmitter (transmitter *B* in the figure), is connected to a dummy load. The operation of the transmitters is continually monitored, and if the working transmitter fails, the transmitter RF switch switches the standby transmitter to the antenna duplexer, thus restoring transmission almost instantly. On the receive side, the RF input signal is

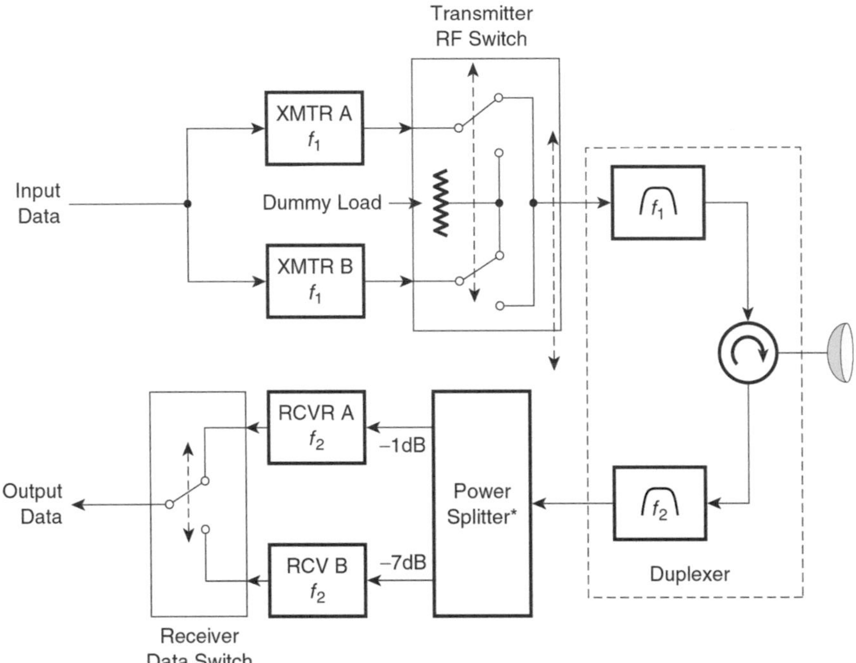

Fig. 7.10 Single-channel monitored hot standby terminal (one way, for simplicity)

normally unevenly split by a power splitter. Typically, the splitter imparts a loss of about 1 dB in the path of the working receiver (receiver A in the figure), i.e., the receiver normally connected to the data output port, and a loss of about 7 dB in the path of the standby receiver. Should the working receiver fail, then the receiver data switch switches to the standby receiver, receiver B in the figure, restoring transmission almost instantly, as in the case of a transmitter failure. When operating on receiver B, the received signal is attenuated by 6 dB relative to when operated on receiver A. However, as the operation on receiver B is rare, this penalty is deemed acceptable, versus the use of an equal loss splitter that would impart about 3.5 dB loss to each receiver, and thus an additional 2.5 dB loss to path A relative to the 1/7 dB splitter arrangement.

Monitored cold standby protection is an alternative to monitored hot standby protection. Here, the difference to hot standby is that, with reference to Fig. 7.10, both the standby transmitter, transmitter B, and the standby receiver, receiver B, are normally not turned on for operation, i.e., "cold." Further, there is no need for the dummy load on the transmit side. Failure of the normally operating transmitter or receiver results in a restoration of transmission via the standby component, but loss of data is larger than in the hot standby case as the standby component will take some finite time to power up to normal operation. The advantage here is savings in operating power.

7.2.6 The Antenna

Antennas used in point-to-point wireless transport links have traditionally been parabolic, but flat-panel versions have begun to be employed. Both such antennas were described in Sect. 3.2. For the parabolic antenna, we note that it comes in both single as well as dual-polarization configurations. Further, in support of band aggregation, for example, one link transmitting in both the E-band and the 23 GHz band, effective multiband antennas using a single dish are now available (see Fig. 7.3).

The gain of antennas used on wireless transport links typically fall in the range of 30–50 dBi, and this gain range corresponds to half-power beamwidth of 5–0.5 degrees, respectively [2]. Those antennas at the higher end of the gain range tend to be utilized in the E-band where the average gain is about 48 dBi [2]. The result is that extra attention needs to be paid to antenna alignment when such high gain antennas are deployed. Further, even if properly aligned during installation, mast sway due to wind loading can temporarily de-align such antennas. A solution to this problem is the use of mast sway compensation technology which dynamically compensates for the tower sway.

7.3 Link Capacity Capability

The capacity of a single channel, single polarized 2^n- QAM modulated wireless link, with no LoS MIMO, is a function of:

- The channel bandwidth, *CBW*
- QAM modulation spectral efficiency, n
- The RF signal roll-off factor, α, where α is the CBW divided by the RF signal Nyquist
- bandwidth
- Coding loss, *CL*, defined as the user data rate divided by the user data plus coding overhead rate

We define the capacity of the terminal post header and payload compression as the *air interface capacity* (AIC). It is easily shown that *AIC* is given by:

$$AIC = CBW \times n \times \frac{1}{\alpha} \times CL \tag{7.1}$$

For the traditional bands:

Table 7.1 Typical air interface capacity (Mb/s) in traditional bands

Modulation	Channel bandwidth (CBW)		
	28 MHz	56 MHz	112 MHz
256-QAM	175	350	700
512-QAM	197	394	787
1024-QAM	219	437	875
2048-QAM	241	481	962
4096-QAM	262	525	1050
8192-QAM	284	–	–

Table 7.2 Typical air interface capacity (Gb/s) in the E-band

Modulation	Channel bandwidth (CBW)					
	0.125 GHz	0.25 GHz	0.5 GHz	1.0 GHz	1.5 GHz	2.0 GHz
64-QAM	0.56	1.12	2.25	4.5	6.75	9.00
128-QAM	0.66	1.31	2.62	5.25	7.87	10.5
256-QAM	0.75	1.50	3.00	6.00	9.00 (future)	12 (future)
512-QAM	0.84	1.69	3.37	–	–	–

- The maximum channel *CBW*s vary from about 28 MHz to 112 MHz.
- The modulations used typically vary from BPSK up to a maximum of 8196-QAM.
- The roll-off factor, α, is typically on the order of 1.15–1.2.
- Coding overhead is typically no more than about 10%; thus coding loss, *CL*, is typically no more than about $1/1.1 \simeq 0.9$.

Table 7.1 shows approximate *AIC* likely in traditional bands for various modulations, a roll-off factor of 1.15, and a coding loss of 0.9.

For the E-band:

- The maximum channel *CBW*s vary from about 125 MHz to 2 GHz.
- The modulations used typically vary from BPSK up to a maximum of 512-QAM.
- The roll-off factor, α, is typically on the order of 1.2.
- Coding overhead, is typically no more than about 10%; thus coding loss, *CL*, is typically no more than about $1/1.1 \simeq 0.9$.

Table 7.2 shows approximate *AIC* likely in the E-band for various modulations, a roll-off factor of 1.2, and coding loss of 0.9.

We note that in the E-band, for a channel bandwidth of 2 GHz, the current typical maximum modulation order is 128-QAM (2^7-QAM). At first glance, this seems to be a missed opportunity to increase capacity by increasing modulation order. Why not use say 2048-QAM (2^{11}-QAM)? This would result in an increase in capacity of $11/7 = 1.57$, i.e., a 57% increase. The answer lies in two limiting factors. The first and the more important is phase noise which, despite current efforts to minimize its influence, limits performance due to a) its intrinsic high value at the operating frequencies in this band and b) its impact on highly compacted signal states at a very high modulation order. The second limiting factor, given the much higher modulation order, is the 12 dB increase in the 10^{-6} BER threshold level that would be incurred even with effective phase noise mitigation. The end result of this would be, without a similar increase in received signal level via increased transmit power and or antenna gain, a much smaller fade margin and hence a much lower reliability for a given path length or a much shorter path length for a given reliability.

As previously stated, the *AIC* above is the capacity of a single-channel 2^n- QAM modulated wireless link, with single polarization and no LoS MIMO. Letting:

- Cross-polarization gain, *XP*, be given by $XP = 1$ for no cross-polarization, and $XP = 2$ for cross-polarization
- MIMO gain be *N* for *N*x*N* spatial MIMO

Then, full single-channel air interface capacity, AIC_F, is given by:

$$AIC_F = AIC \times XP \times N \tag{7.2}$$

Thus, for transmission in a traditional band, with a *CBW* of 112 MHz and 4096-QAM modulation where the *AIC* is approximately 1 Gb/s, then when also employing cross-polarization as well as 2×2 MIMO, AIF_F computes to approximately 4 Gb/s.

For transmission in the E-band, with a *CBW* of 2 GHz and 128-QAM modulation where the *AIC* is approximately 10 Gb/s, then when also employing cross-polarization as well as 2×2 MIMO, AIF_F computes to approximately 40 Gb/s!

Recall from above that we defined the *air interface capacity*, *AIC*, as the capacity of the Ethernet conveyed signal that has already undergone header and payload compression. Thus, the capacity that the link can handle prior to header and payload compression is larger than the air interface capacity as defined. This increase in capacity is a function of the Ethernet frame size. The Ethernet header and trailer total 18 Bytes and an IPv4 header is 20 bytes. Thus, for the smallest frame size (64 bytes) header compression is huge. For the largest frame size, a jumbo frame as long as 9216 bytes, header compression is essentially zero. As the frame size varies throughout transmission it is not possible to put a fixed number on the capacity improvement as a result of header and payload compression as well as the elimination of the Ethernet preamble, start frame delimiter, and interframe gap. However, an overall average capacity gain of somewhere in the region of 10% as a result of compression and elimination is not unreasonable.

To demonstrate the potential of E-band radios in combination with LoS MIMO, Deutsche Telekom and Ericsson jointly trialed a very high-capacity wireless system in Athens, Greece [2]. The salient features of the link were as follows:

- Link length: 1.5 km
- Operating frequency: 73 GHz
- Channel bandwidth: 2 GHz and 2.5 GHz
- Modulation: 64, 128, and 256-QAM
- Cross-polarized transmission

- 4×4 spatial multiplexing with the 4 antennas at each end located on the 4 corners
  of a 1.7 m $\times$ 1.7 m square
- Individual antenna: 0.6 m parabolic reflector

Assuming a roll-off factor, α, of 1.2 and a coding loss, *CL* of 0.9, then by Eq. 7.1, the *AIC* with the 2.5 GHz channel bandwidth computes to 11.25, 13.1, and 15 Gb/s for 64, 128, and 256-QAM modulation respectively. By Eq. 7.2, the associated full link capacity, AIC_F, computes to 90, 105, and 120 Gb/s. The actual full link capacity achieved was stated to be 105, 126, and 139 Gb/s, highly impressive numbers indeed. The difference between the actual and those calculated is likely due to different actual values of α and *CL*, as well as overhead compression and elimination gain. The link had a fade margin of better than 25 dB resulting in a rain-limited availability of better than 99.99%. It demonstrated sub 5 µs latency at 100 Gb/s.

7.4   Example Specifications and Typical Path Performance of an 80 GHz (E-Band) Link

To gain an insight into the specifications and path performance of real wireless terminals used for mobile network transport, we will look at two such terminals. First, in this section, we will review an Ericsson terminal that operates in a non-traditional band. In the following section, we will review an Ericsson terminal that operates in a traditional band.

The MINI-LINK 6352 operates in the E-band and affords extremely high transmission rates. It is an all-outdoor microwave node, employs direct conversion and its simplified block diagram is thus somewhat similar to that shown in Fig. 7.1a. Photos of the 6352 are shown in Figs. 7.3 and 7.11.

Stated below, courtesy of Ericsson, are some example specifications:

- Operating frequency range: 71–76/81–86 GHz
- Channel bandwidths (CBW): 125, 250, 500, 750, 1000, 1500, and 2000 MHz
- Modulation supported:

  - BPSK to 512-QAM for CBWs of 125–500 MHz
  - BPSK to 256-QAM for CBWs of 750–1000 MHz
  - BPSK to 128-QAM for CBWs of 1500–2000 MHz

- Air interface capacity, single channel, single polarized:

  - 9.6 Gb/s for 2000 MHz CBW and 128-QAM
  - 6.0 Gb/s for 1000 MHz CBW and 256-QAM
  - 3.4 Gb/s for 500 MHz CBW and 512-QAM

Fig. 7.11 Photo of the Ericsson MINI-LINK 6352. (With the permission of Ericsson)

- Antennas for integrated installation:

  - 0.1, 0.2, 0.3, 0.6 m, single polarized
  - 0.3 and 0.6 m, dual polarized

- CCDP transmission and XPIC: Supported
- Header compression: Multilayer header compression, enabling as high as 20% extra capacity
- Transmitter power:

  - + 16 dBm for 128-QAM
  - + 15 dBm for 256- and 512-QAM

- Examples of receiver sensitivity (BER 10^{-6}):

  - − 48 dBm for 500 MHz CBW, 512-QAM
  - − 48 dBm for 1000 MHz CBW, 256-QAM
  - − 48 dBm for 2000 MHz CBW, 128-QAM

- Coding: LDCP and Reed Solomon.
- Adaptive modulation and coding: Supported

Table 7.3 Availability versus path length for assumed 80 GHz link with 0.6 m antennas

Availability (%)	99.999	99.99	99.9
Path length (km)	1.4	2.6	5.6

– Ethernet frame size: 64–1518 bytes, standard, 9216 Jumbo.
– Ethernet traffic interfaces:

  • 3 × SFP+ optical port, each configurable to either 1, 2.5, and 10 Gb/s
  • 1 × 100/1000Base-T electrical port, supporting Power over Internet (PoE)

– Configurations:

  • 1 + 0, i.e., 1 channel, unprotected
  • 2 + 0 Radio link bonding, i.e., 2 channels, unprotected, with or without XPIC
  • 1 + 1 Cold standby, i.e., 1 channel with another in cold standby mode to take
     over in case of failure   To get a sense of the availability afforded by a link
  using the MINI-LINK 6352, consider a link with the following parameters:

– Location: Central Europe, latitude 50^0 north, 30 mm/h rain rate $R_{0.01}$ exceeded
   0.01% of time
– Average of go-and-return transmission frequency: 78 GHz
– Polarization: Vertical
– Channel bandwidth: 2000 MHz
– Modulation: 128-QAM
– Air Interface Capacity: 9.6 Gb/s
– Antennas: 0.6 m at each end, each with a gain, G_a, of 51.0 dB
– Transmit power, $P_t = +16$ dBm
– Receiver sensitivity, $R_{sens} = -48$ dBm

For a path length of d km, path loss, L_p, computes to $(130.2 + 20\log_{10}d)$ dB via
Eq. 3.11a. The received power, P_r, would thus be given by

$$P_r = P_t + (2 \times G_a) - L_p$$
$$= +16 + (2 \times 51.0) - 130.2 - 20\ \log_{10}d$$
$$= (-12.2 - 20\log_{10}d)\ \text{dBm} \tag{7.3}$$

And the fade margin FM would be given by

$$FM = P_r - R_{sens} = (35.8 - 20\log_{10}d)\ \text{dB} \tag{7.4}$$

It can be shown, using the process outlined in Appendix C, that for a path of
length 1.4 km the rain attenuation exceeded 0.001% of the time is 33 dB which is
also the fade margin for this path length. The path thus has an availability of
99.999%. It can similarly be shown that a path of length 2.6 km has an availability
of 99.99% and a path of length 5.6 km has an availability of 99.9%. This relationship
is summarized in Table 7.3.

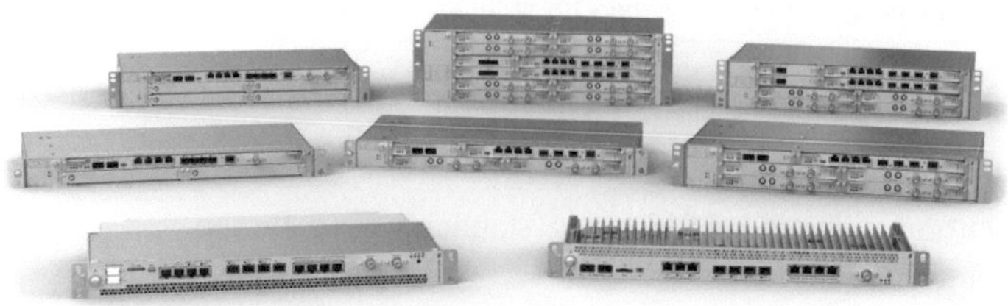

Fig. 7.12 Photo of many options of the Ericsson MINI-LINK 6600. (With the permission of Ericsson)

Fig. 7.13 Photo of the Ericsson MINI-LINK 6363. (With the permission of Ericsson)

7.5   Example Specifications and Typical Path Performance of a 32 GHz Link

The Ericsson MINI-LINK 6600/6363 combination operates in the traditional bands (6–42 GHz). However, here, we will focus on operation in the 32 GHz band. The 6600 is in effect an indoor baseband and IF processor, and the 6363 is an outdoor heterodyne RF front end. A block diagram of the combination is thus somewhat similar to that shown in Fig. 7.1b. A photo of many options of the 6600 is shown in Fig. 7.12 and of the 6363 in Figs. 7.3 and 7.13.

Stated below, courtesy of Ericsson, are some example specifications for the 32 GHz option:

- Operating frequency range: 31.8–33.4 GHz
- Channel bandwidths (CBW): 7, 14, 28, 56, and 112 MHz
- Modulation supported:

- 4- to 4096-QAM for CBWs of 28, 56, and 112 MHz
- 8192-QAM supported for CBW of 28 MHz and adaptive modulation

– Air interface capacity, single channel, single polarized, 4096-QAM:

- 264 mb/s for 28 MHz CBW
- 534 Mb/s for 56 MHz CBW
- 1068 Mb/s for 112 MHz CBW

– Antennas:

- 0.2 to 0.9 m, single and dual polarized
- 0.1 M flat panel

– CCDP/XPIC: Supported for most modes.
– LoS 2 × 2 MIMO: Supported for 28 MHz CBW up to 512-QAM and for 56 MHz CBW up to 1024-QAM
– Header compression: Multilayer header compression supported
– Transmitter power:

- + 17 dBm for 4096-QAM

– Example receiver sensitivity (BER 10^{-6}), 4096-QAM:

- − 54.5 dBm for 28 MHz CBW
- − 51.5 dBm for 56 MHz CBW
- − 48.5 dBm for 112 MHz CBW

– Coding: LDCP and Reed Solomon
– Adaptive modulation and coding: Supported
– Ethernet frame size: 64–1518 bytes, standard, 9216 Jumbo
– Ethernet traffic interface types:

- 1000BASE-SX/LX/ZX/BX, optical
- 1000BASE-X CWDM, optical
- 10GBASE-LR/ER/ZR, optical
- 10BASE-T/100BASE-TX/1000BASE-T, electrical

– Configurations:

- 1 + 0, i.e., 1 channel, unprotected
- 2 + 0, i.e., 2 channels, unprotected
- 4 + 0, i.e., 4 channels, unprotected
- 1 + 1 Hot standby, i.e., 1 active channel with another in hot standby mode ready to take over in case of failure
- 2 + 2 Hot standby, i.e., 2 active channels with another two in hot standby mode ready to take over in case of failure   To get a sense of the availability afforded by a link using the MINI-LINK 6600/6363 combination, consider a link with the following parameters:

Table 7.4 Availability versus path length for assumed 32 GHz link with 0.6 m antennas

Availability (%)	99.999	99.99	99.9
Path length (km)	2.0	3.6	7.2

Table 7.5 Availability versus path length for assumed 32 GHz link with 0.9 m antennas

Availability (%)	99.999	99.99	99.9
Path length (km)	2.5	4.8	10.2

- Location: Central Europe, latitude 500 north, 30 mm/h rain rate R0.01 exceeded 0.01% of time 30
- Average of go-and-return transmission frequency: 32.6 GHz
- Polarization: Vertical
- Channel bandwidth: 112 MHz channel
- Modulation: 4096-QAM
- Air interface capacity: ~ 1.1 Gb/s
- Antennas: 0.6 m at each end, each with a gain, Ga, of 43.8 dB
- Transmit power, Pt = + 17 dBm
- Receiver sensitivity, Rsens = −48.5 dBm

For a path length of d km, path loss, L_p, computes to $(122.5 + 20\log_{10}d)$ dB via Eq. (3.11a). The received power, P_r, would thus be given by

$$
\begin{aligned}
P_r &= P_t + (2 \times G_a) - L_p \\
&= +17 + (2 \times 43.8) - 122.5 - 20\ \log_{10}d \\
&= (-17.9 - 20\log_{10}d)\ \text{dBm}
\end{aligned}
\tag{7.5}
$$

And the fade margin FM would be given by

$$
FM = P_r - R_{sens} = (30.6 - 20\log_{10}d)\ \text{dB}
\tag{7.6}
$$

It can be shown, using the process outlined in Appendix C, that for a path of length 2.0 km the rain attenuation exceeded 0.001% of the time is 24.5 dB which is also the fade margin for this path length. The path thus has an availability of 99.999%. It can similarly be shown that a path of length 3.6 km has an availability of 99.99% and a path of length 7.2 km has an availability of 99.9%. This relationship is summarized in Table 7.4.

Had we chosen to assume 0.9 m antennas each with a gain of 47.8 dB, such antennas being the largest typically used in this frequency band, then the fade margin would have increased by 8 dB relative to the use of 0.6 m antennas and the path lengths for a given availability would be increased as shown in Table 7.5.

We note that the availability analyses of both this 32 GHz link and E-band one outlined in Sect 7.4 above assume the same location and same rain conditions. Thus, if we assume that the 32 GHz transmission uses 0.9 m antennas and the two transmissions are aggregated over a common 2.5 km path, then we would have a composite link with a total capacity of approximately 10.7 Gb/s, a reliability >99.9% for this total capacity, and a reliability of 99.999% for a capacity of 1.1 Gb/s.

7.6 Conclusion

In Chap. 1, the overall architecture of 5G mobile communications was reviewed. Usage scenarios and top-level requirements were addressed, and the place in the architecture of transport systems, specifically backhaul, midhaul, and fronthaul, was elucidated. The role of wireless in the realization of these transport systems was outlined, and the key wireless transport technologies employed in such system realization were summarized. In Chap. 2, Ethernet-based packet-switched data was described, this being the form of data carried by 5G mobile network transport links. Chap. 3 outlined the nature of the wireless path traversed by such transport links and the impact of path characteristics on path availability.

The information presented in Chaps. 1, 2 and 3, though somewhat peripheral, was important to truly understand the need for and the necessary capabilities of the key technologies employed in the realization of wireless transport links. Chapters 4, 5 and 6 addressed these technologies. Information was provided at a level sufficient to impart a fundamental grasp of the structure and functioning of these technology components, but not at a level so deep as to make it somewhat intractable to those with a limited background in the subject. Chapter 4 dealt with QAM modulation, a key driver to higher spectral efficiency. Chapter 5 dealt with techniques that optimize performance such as forward error correction, quadrature modulation/demodulation imperfection mitigation, and adaptive equalization of path imperfections. Chapter 6 dealt with non-modulation-based capacity improvement techniques such as co-channel dual polarization transmission/cross polarization interference cancellation and line-of-sight MIMO. Finally, in this chapter, the block-level architecture of typical mobile wireless transport terminals was outlined, link capacity capability reviewed, and key specifications and typical path performance of two commercial terminals presented.

Fixed wireless technology is ever-changing and ever-improving. This improvement comes, however, at the expense of ever more complexity. There is no reason to believe that many as yet unproven technologies will not emerge and supplement or replace those in current use. That said, a lot of technologies presented in this text will likely have a long shelf life. Take digital modulation. Many variations exist, but the fundamental property of allowing an increase in spectral efficiency at the expense of poorer signal-to-noise characteristics is unchanging. The same long shelf-life comment applies equally to channel coding. Yes, it's true that no sooner is a particular type of coding crowned as the ultimate that a new one emerges. In the last decade, turbo convolution coding has been supplanted by low-density parity check (LDPC) coding which is now almost universally used in fixed wireless transport links. However, it is possible that in the future polar codes, which have found a role in

5G mobile access, may find a role in wireless transport systems. Hence, its introduction in Chap. 5. LoS MIMO has found its place in fixed wireless as a reliable means of increasing capacity. Orbital angular momentum (OAM) multiplexing, on the other hand, is still very much in the research phase. However, if development leads to practical implementation, then, like LoS MIMO, it will be a useful tool in the quest for ever-increasing capacity. It is hoped that having studied this introduction to the key technologies underlying wireless transport, should the reader desire to explore this subject area in greater detail, he/she will feel well-positioned to do so.

References

1. Camarchi V et al (2016) Electronics for microwave backhaul. Artech House, Norwood
2. Ericsson (2019) Ericsson Microwave Outlook; October 2019, Goteborg, Sweden

Appendices

Appendix A

Helpful Mathematical Identities

Trigonometric Identities

$$\sin(x \pm y) = \sin x \cos y \pm \cos x \sin y$$

$$\cos(x \pm y) = \cos x \cos y \pm \sin x \sin y$$

$$\sin x \sin y = \frac{1}{2}\cos(x - y) - \frac{1}{2}\cos(x + y)$$

$$\cos x \cos y = \frac{1}{2}\cos(x + y) + \frac{1}{2}\cos(x - y)$$

$$\sin x \cos y = \frac{1}{2}\sin(x + y) + \frac{1}{2}\sin(x - y)$$

$$\cos x \sin y = \frac{1}{2}\sin(x + y) - \frac{1}{2}\sin(x - y)$$

$$\sin^2 x = \frac{1}{2}(1 - \cos 2x)$$

$$\cos^2 x = \frac{1}{2}(1 + \cos 2x)$$

$$\sin x = \frac{e^{jx} - e^{-jx}}{2j}$$

$$\cos x = \frac{e^{jx} + e^{-jx}}{2}$$

$$e^{jx} = \cos x + j \sin x$$

Standard Integrals

Where a, b, and c are constants,

$$\int \sin (ax + b)\ dx = -\frac{1}{a} \cos (ax + b) + c$$

$$\int \cos (ax + b)dx = \frac{1}{a} \sin (ax + b) + c$$

$$\int adx = ax + b$$

$$\int (ax + b)^n dx = \frac{1}{a(n + 1)} (ax + b)^{n+1}\ \ n \neq -1$$

$$\int e^{ax+b}dx = \frac{1}{a}e^{ax+b} + c$$

Matrix Algebra

Matrix product example:

$$\overset{3\,x\,2}{\begin{pmatrix} a_{11} & a_{12} \\ a_{21} & a_{22} \\ a_{31} & a_{32} \end{pmatrix}} \overset{2\,x2}{\begin{pmatrix} b_{11} & b_{12} \\ b_{21} & b_{22} \end{pmatrix}} = \overset{3\,x\,2}{\begin{pmatrix} a_{11}b_{11} + a_{12}b_{21} & a_{11}b_{12} + a_{12}b_{22} \\ a_{21}b_{11} + a_{22}b_{21} & a_{21}b_{12} + a_{22}b_{22} \\ a_{31}b_{11} + a_{32}b_{21} & a_{31}b_{12} + a_{32}b_{22} \end{pmatrix}}$$

Appendix B

Multipath Fading Outage Analysis

B.1 Total Outage

Following is a method to predict *total outage* on an unprotected digital link affected by multipath fading and thermal noise. It is based on ITU recommendation ITU-R P.530-17 [1] and is applicable to "quick planning applications."

For outage on a single polarized unprotected digital link, total outage P_t is given by

$$P_t = P_{ns} + P_s \tag{B.1}$$

where P_{ns} is the outage due to the nonselective component of the fading and is in fact that outage due to error rate exceeding a given threshold as a result of decreased signal-to-noise ratio and is thus related to the flat fade margin

and P_s is the outage due to the selective component of the fading, is independent of the signal-to-noise ratio, and can be viewed as the outage that would result if the fade margin was infinite.

B.2 Unprotected Nonselective Outage Prediction

Probability of nonselective outage P_{ns} is given by

$$P_{ns} = p_w/100 \tag{B.2}$$

where p_w is the percentage of time that the flat fade margin A (dB) corresponding to a specified bit error rate (BER) is exceeded in the average worst month. The flat fade margin is the difference between the unfaded received signal level and the receiver threshold for the specified BER and calculated via specifications of the terminal equipment and the path data.

The percentage p_w is given by

$$p_w = Kd^{3.1}\left(1 + |\varepsilon_p|\right)^{-1.29} f^{0.8} \times 10^{-0.00089h_L - A/10} \tag{B.3}$$

where

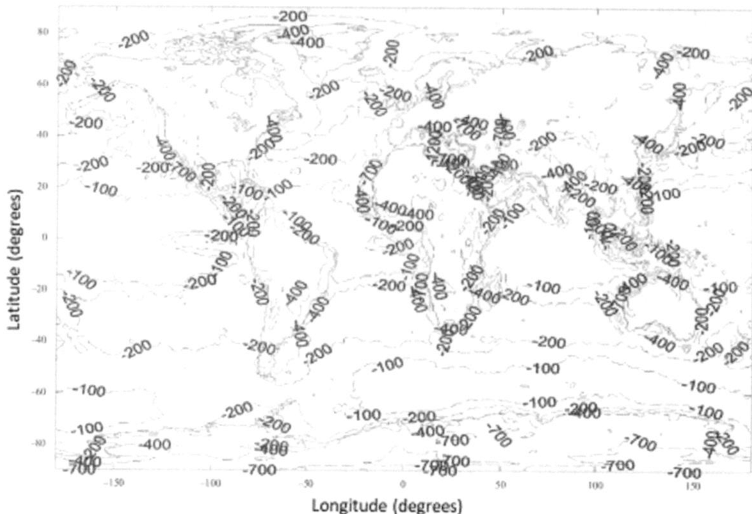

Fig. B.1 Refractivity gradient dN_1 not exceeded for 1% of the average year on the lowest 65 m. (From [2], with the permission of the ITU)

$$K = 10^{-4.6 - 0.0027 dN_1} \qquad (B.4)$$

and d is the path length in km

    f is the frequency in GHz

    dN_1 can be obtained from Fig. B.1 [2].

and

$$|\varepsilon_p| = |h_r - h_e|/d \qquad (B.5)$$

where h_e and h_r are the antenna heights in meters above sea level

    and h_L is the altitude of the lower antenna.

B.3 Unprotected Selective Outage Prediction

The probability of *selective outage P_s* is given by

$$P_s = 2.15\eta \left(\left(W_M \times 10^{-B_M/20} \times \frac{\tau_m^2}{\tau_{r,M}} \right) + \left(W_{NM} \times 10^{-B_{NM}/20} \times \frac{\tau_m^2}{\tau_{r,NM}} \right) \right) \qquad (B.6)$$

where:

W_x is signature width (GHz),
B_x is signature depth (dB),
$\tau_{r,x}$ is the reference delay (ns) used to obtain the signature,
subscript x denotes either minimum phase (M) or non-minimum phase (NM) fades,
τ_m is the mean time delay,
η is the multipath activity parameter.

τ_m being given by

$$\tau_m = 0.7\left(\frac{d}{50}\right)^{1.3} \text{ ns} \tag{B.7}$$

and η being given by

$$\eta = 1 - e^{-0.2(P_0)^{0.75}} \tag{B.8}$$

where $P_0 = p_w/100$ and p_w is calculated from Eq. (B.3) with $A = 0$ dB.

B.4 Outage Prediction Example

Consider a radio transceiver with the following commercially realizable specifications:

Operating frequency: 6 GHz
Channel bandwidth: 28 MHz
Modulation: 4096-QAM
Air interface capacity: ~ 260 Mb/s
Transmitter output power: +28 dBm
Receiver 10^{-6} BER threshold: -55 dBm typical
Signature width $W_M = 31$ MHz
Signature width $W_{NM} = 32$ MHz
Signature depth $B_M = 25$ dB for 10^{-6} BER
Signature depth $B_{NM} = 24$ dB for 10^{-6} BER
Signature reference delay $\tau_{r,M} = \tau_{r,NM}$: 6.3 ns

Consider that this transceiver is operating over a path with the following parameters:

Path length: 20 km
Antennas: Each 3.7 m (12 ft.) diameter with gain of 44.5 dB

Antenna height h_r: 300 m
Antenna height h_e: 287 m
Path location: Northern India where $dN_1 = -400$
Path clearance above closest obstruction such that path suffers no diffraction loss

Probability of nonselective outage computation
    With the above data, we can compute that:

Free space loss = 134 dB
Received signal level = $+28 + 44.5 - 134 + 44.5 = -17$ dBm
Fade margin = $-17 - (-55) = 38$ dB = A dB
$K = 0.0003$ (Eq. B.4)

$$|e_p| = |300 - 287|/20 = 0.65$$

and hence, the probability of nonselective outage p_w, by Eq. B.3, is given by

$$p_w = 0.0003 \times 20^{3.1} \times (1 + 0.65)^{-1.29} \times 6^{0.8} \times 10^{-(0.249+(38/10))} = 0.00041\%$$

and thus, $P_{ns} = 0.0000041$.

Probability of selective outage computation

With the above data, we can compute that:
The mean time delay $\tau_m = 0.213$ ns
The percentage of time p_w that the fade depth 0 dB is exceeded in average worst
    month = 50 (Eq. B.3) and hence $P_0 = 0.5$
Activity parameter $\eta = 0.112$ (Eq. B.8)

    and hence, the probability of selective fading P_s, by Eq. B.6, is given by

$$P_s = 2.15 \times 0.112 \left(\left(0.031 \times 10^{\frac{25}{20}} \times \frac{0.213^2}{6.3} \right) + \left(0.032 \times 10^{\frac{24}{20}} \times \frac{0.213^2}{6.3} \right) \right)$$

$$= 0.0000065$$

Total outage probability

By B.1, total outage probability $P_t = P_{ns} + P_s = 0.0000041 + 0.0000065 = 0.0000106$.

Thus, as a percentage, total outage probability = 0.00106%, and hence, path availability is approximately 99.999%.

References

[1]. ITU (2017) Recommendation ITU-R P.530-17: propagation data and prediction methods required for the design of terrestrial line-of-sight systems. ITU, Geneva
[2]. ITU (2012) Recommendation ITU-R P.453-10: the radio refractive index; its formula and refractivity data. ITU, Geneva

Appendix C

Rain Outage Analysis

To determine the fade margin required for a given rain-related path availability requires knowledge of the rain-induced path attenuation exceeded for a given percentage of the time. The latter, in turn, requires knowledge of the effective path length and the probability distribution of rain attenuation as a function of frequency, polarization in the general geographic vicinity of the path. ITU Recommendation ITU-R P.837-7 [1] gives a map of the world that's color-graded as a function of the rain rate, $R_{0.01}$, that's exceeded 0.01% of the time. ITU-R P.838-3 [2] gives a procedure to determine the specific attenuation γ_R (dB/km), and ITU-R P.530-10 [3] gives a formula to determine the effective path length d_{eff}. A procedure to use the data provided in these recommendations to compute the path attenuation $A_{0.01}$ exceeded 0.01% of the time on a non-cross-polarized path is as follows:

Step 1: Obtain $R_{0.01}$ from Fig. C.1 (Figure 1 of [1]).
Step 2: From [2], determine the specific attenuation γ_R using the equation.

$$\gamma_R = k(R_{0.01})^{\alpha} \tag{C.1}$$

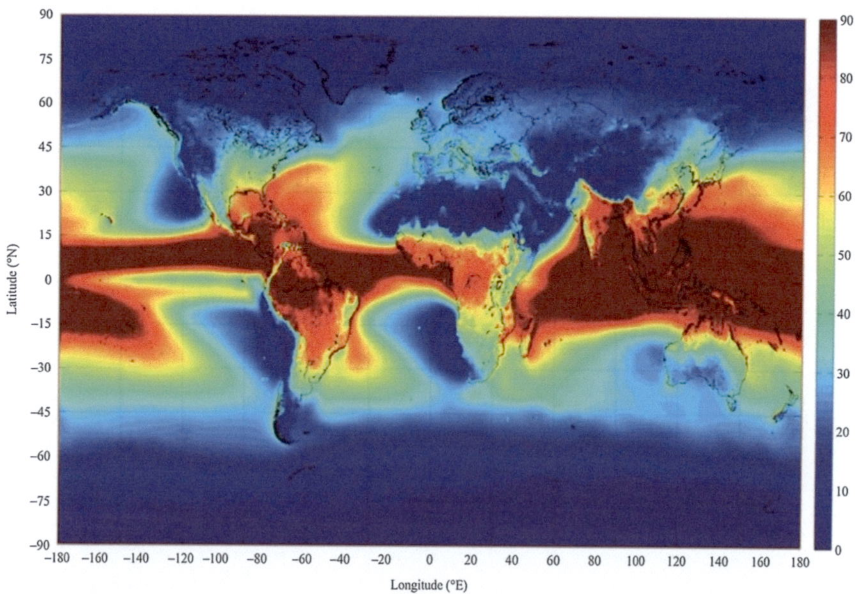

Fig. C.1 Rainfall rate exceeded 0.01% of an average year. (From [1], with the permission of the ITU)

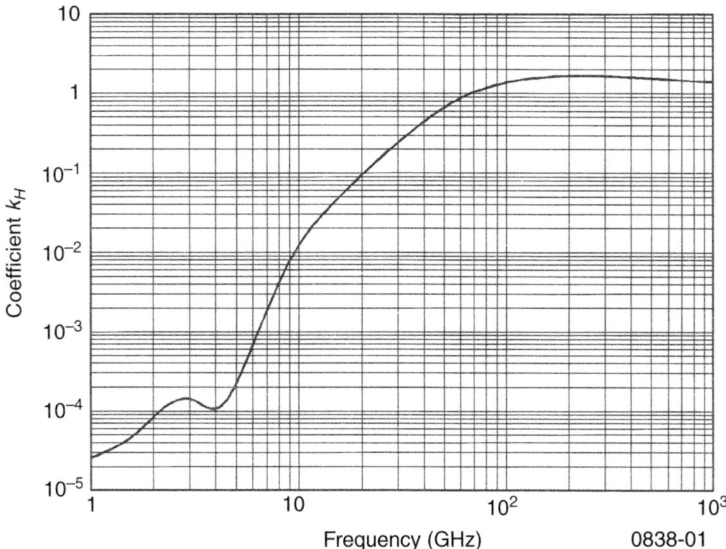

Fig. C.2 *k* Coefficient for horizontal polarization. (From [2], with the permission of the ITU)

where the values of the coefficients k and α are a function of frequency and polarization and k can be determined, depending on the polarization, from Fig. C.2 or C.3, and α can be determined, depending on the polarization, from Fig. C.4 or C.5. (Note that an alternate but somewhat less accurate way to determine γ_R is to estimate if from Fig. 3.15)

Step 3: Compute the *effective path distance* d_{eff} of the link. The effective path distance can be less than the actual path length because rain cells are of limited dimension and varying intensity, and thus, don't necessarily cover the full path at a fixed rate. The effective path distance d_{eff} is given by [3] to be:

$$d_{eff} = d\,\frac{1}{1 + d/d_0} \qquad\qquad (C.2)$$

where d is the actual path length in km
and where, for $R_{0.01} \leq 100$ mm/h,

$$d_0 = 35e^{-0.015R_{0.01}} \qquad\qquad (C.3)$$

and where, for $R_{0.01} > 100$ mm/h, use the value 100 mm/h in place of $R_{0.01}$.

Figure C.6 shows effective path distance d_{eff} for rain rates of 30, 60, and 90 mm/h. A rain rate of or close to 30 mm/h is typical of large parts of Europe and Eastern China. A rain rate of or close to 60 mm/h is typical of parts of the South Eastern United States, Northern Mexico, and Eastern China. A rain rate of or about 90 mm/h

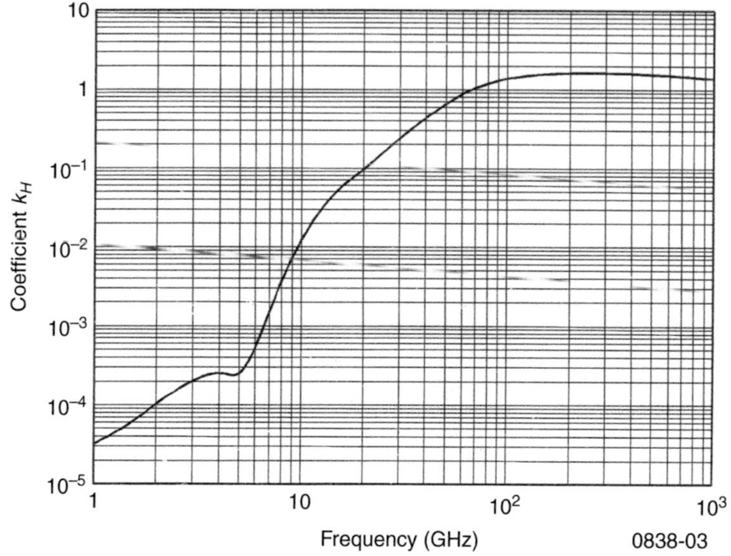

Fig. C.3 k Coefficient for vertical polarization. (From [2], with the permission of the ITU)

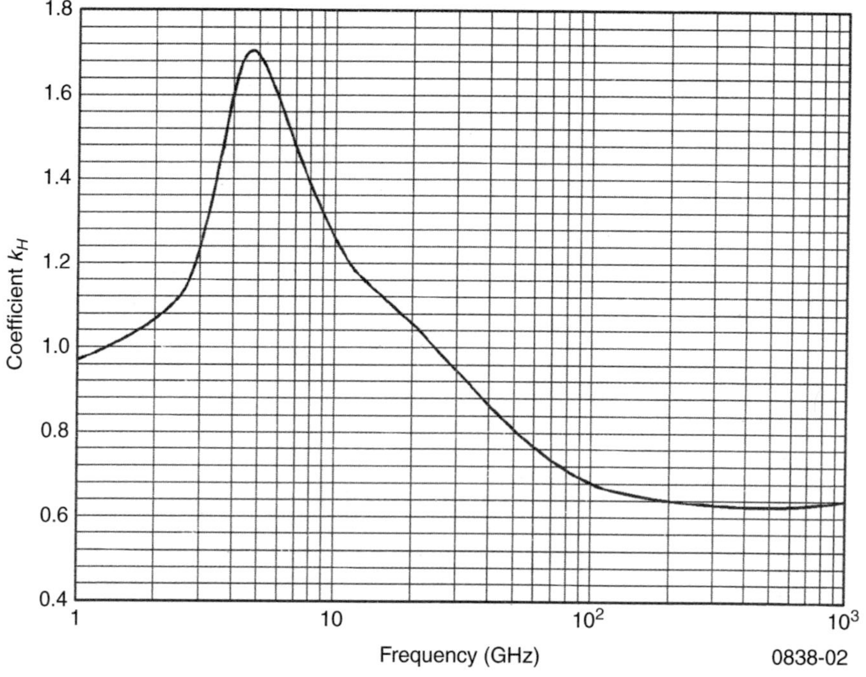

Fig. C.4 α Coefficient for horizontal polarization. (From [2], with the permission of the ITU)

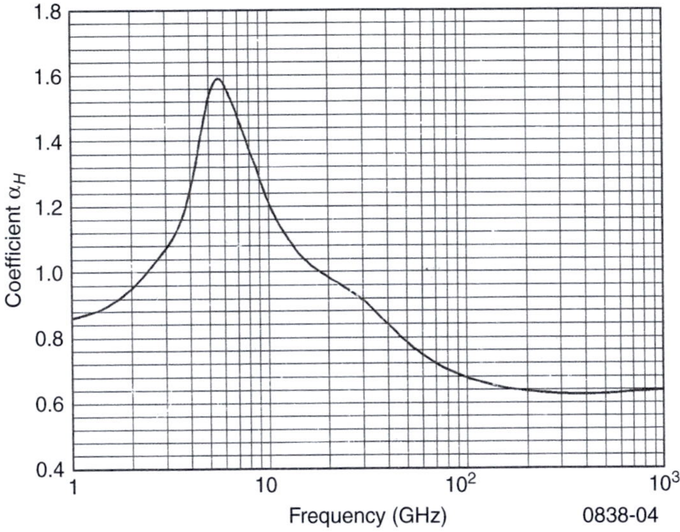

Fig. C.5 α Coefficient for vertical polarization. (From [2], with the permission of the ITU)

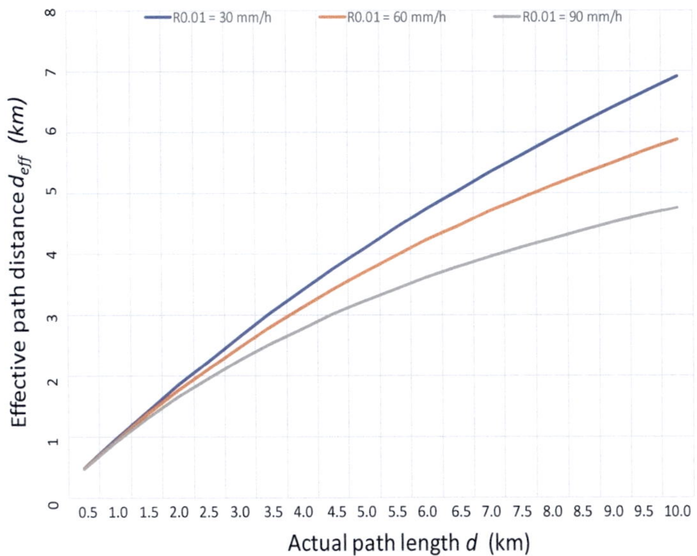

Fig. C.6 Effective path distance for certain rain rates

is typical of parts of southern India, South East Asia, and the northern part of South America

Step 4: Compute $A_{0.01}$, the path attenuation due to rain exceeded 0.01% of the time, via the relationship

Table C.1 Factor F that $A_{0.01}$ must be multiplied by to determine A_p, the path attenuation exceeded p % of the time

Radio links located in latitudes equal to or greater than 30^0 (north or south)				
p	1%	0.10%	0.01%	0.001%
F	0.12	0.39	1	2.14
Radio links located in latitudes less than 30^0 (north or south)				
p	1%	0.10%	0.01%	0.001%
F	0.07	0.36	1	1.44

$$A_{0.01} = \gamma_R d_{eff} \ \text{dB} \tag{C.4}$$

Note that if the fade margin is equal to $A_{0.01}$, then this results in a rain-related path availability of 99.99% and rain outage events totaling 53 minutes per year.

Step 5: Compute, as per [3], A_p, where A_p is the path attenuation due to rain exceeded p % of the via

$$A_p = A_{0.01} F \tag{C.5}$$

where F is given in Table C.1.

The following example demonstrates the application of this procedure.

Example C.1

Fade margin required to achieve a given rain-related availability:

A 32-GHz, 3 km, vertically polarized path is located in central Europe (latitude 50^0 North). What is the fade margin required to achieve a 99.999% path availability?

Solution

The fade margin to achieve a 99.999% path availability is the rain attenuation not exceeded 99.999% of the time, and hence, the rain attenuation exceeded 0.001% of the time, i.e., $A_{0.001}$. To find $A_{0.001}$, we must first, however, find $A_{0.01}$. We find $A_{0.001}$ following the steps outlined above.

Step 1: Determine $R_{0.01}$, the rain level exceeded 0.01% of the time. This is shown by Fig. C.1 to be approximately 30 mm/h.

Step 2: Determine the specific rain attenuation γ_R for $R_{0.01}$. From Fig. C.3, we determine that the coefficient k equals 0.3 and from Fig. C.5 that the coefficient α equals 0.9. Applying these coefficients to Eq. (C.1), we determine that γ_R equals 6.4 dB/km.

Step 3: Determine d_{eff}, the effective path distance, found from Fig. C.6 to be 2.6 km.

Step 4: Compute $A_{0.01}$, the path attenuation due to rain exceeded 0.01% of the time. By Eq. (C.4), $A_{0.01}$= 6.4 x 2.6 = 16.6 dB.

Step 5: Compute $A_{0.001}$. Taking into account that the path latitude is 50^0 North, then from Table C.1, we determine that factor F is 2.14. Thus, $A_{0.001} = 16.6 \times 2.14 = 35.6$ dB.

References

[1]. ITU Recommendation ITU-R P.837-7 (2017) Characteristics of precipitation for propagation modelling. ITU, Geneva

[2]. ITU Recommendation ITU-R P.838-3 (2005) Specific attenuation model for rain for use in prediction methods. ITU, Geneva

[3]. ITU Recommendation ITU-R P.530-10 (2001) Propagation data and prediction methods required for the design of terrestrial line-of-sight systems. ITU, Geneva

Appendix D

Spectral Analysis of Nonperiodic Functions and Linear System Response

D.1 Spectral Analysis of Nonperiodic Functions

A *nonperiodic function* of time is a function that is nonrepetitive over time. A stream of binary data as typically transmitted by digital communication systems is a stream of nonperiodic functions, each pulse having an equal probability of being one or zero, independent of the value of other pulses in the stream. The analysis of the spectral properties of nonperiodic functions is thus an important component of the study of digital transmission.

A nonperiodic waveform, *v(t)* say, may be represented in terms of its frequency characteristics by the following relationship

$$v(t) = \int_{-\infty}^{\infty} V(f) e^{j2\pi ft} df \tag{D.1}$$

The factor $V(f)$ is the *amplitude spectral density* or the *Fourier transform* of *v(t)*. It is given by

$$V(f) = \int_{-\infty}^{\infty} v(t) e^{-j2\pi ft} dt \tag{D.2}$$

Because $V(f)$ extends from $-\infty$ to $+\infty$, i.e., it exists on both sides of the zero-frequency axis, it is referred to as a *two-sided* spectrum.

An example of the application of the Fourier transform that is useful in the study of digital communications is its use in determining the spectrum of a nonperiodic pulse. Consider a pulse *v(t)* shown in Fig. D.1a, of amplitude *V*, and that extends from $t = -\tau/2$ to $t = \tau/2$. Its Fourier transform, $V(f)$, is given by

$$\begin{aligned} V(f) &= \int_{-\tau/2}^{\tau/2} V e^{-j2\pi ft} dt \\ &= \frac{V}{-j2\pi f} \left[e^{-j2\pi f\tau/2} - e^{j2\pi f\tau/2} \right] \\ &= V\tau \frac{\sin \pi f \tau}{\pi f \tau} \end{aligned} \tag{D.3}$$

The form (sin *x*)/*x* is well-known and referred to as the *sampling function, Sa(x)*. The plot of $V(f)$ is shown in Fig. D.1b. It will be observed that it is a continuous function. This is a common feature of the spectrum of all nonperiodic waveforms. We note also that it has zero crossings at $\pm 1/\tau, \pm 2/\tau, \ldots$.

Fig. D.1 Rectangular pulse
and its spectrum

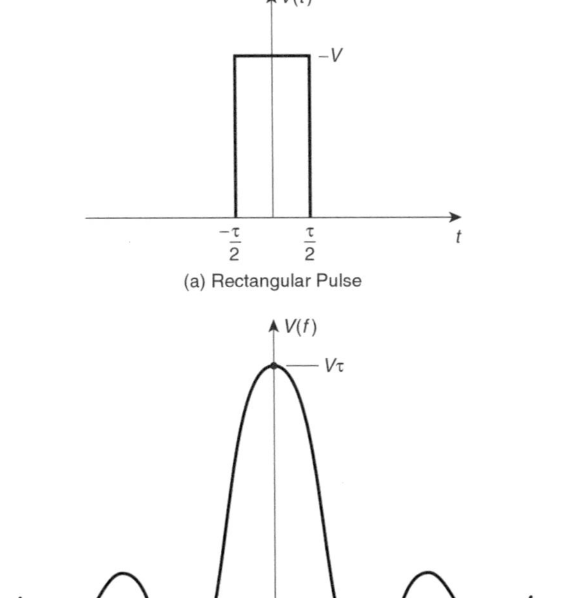

The Fourier transform $V(f)$ of an impulse of unit strength is also a useful result. By definition, an impulse $\delta(t)$ has zero value except at time $t = 0$, and an impulse of unit strength has the property

$$\int_{-\infty}^{\infty} \delta(t)dt = 1 \qquad (D.4)$$

Thus

$$V(f) = \int_{-\infty}^{\infty} \delta(t)e^{-2\pi jft}dt = 1 \qquad (D.5)$$

Equation (D.5) indicates that the spectrum of an impulse $\delta(t)$ has a constant amplitude and phase and extends from $-\infty$ to $+\infty$.

A final example of the use of the Fourier transform is the analysis of what results in the frequency domain when a signal $m(t)$, with Fourier transform $M(f)$, is multiplied by a sinusoidal signal of frequency f_c. In the time domain, the resulting signal is given by

$$v(t) = m(t) . \cos 2\pi f_c t$$

$$= m(t) \left[\frac{e^{j2\pi f_c t} + e^{-j2\pi f_c t}}{2} \right] \tag{D.6}$$

and its Fourier transform is thus

$$V(f) = \frac{1}{2} \int_{-\infty}^{\infty} m(t) e^{-j2\pi(f+f_c)t} dt + \frac{1}{2} \int m(t) e^{-j2\pi(f-f_c)t} dt \tag{D.7}$$

Recognizing that

$$M(f) = \int_{-\infty}^{\infty} m(t) e^{-j2\pi ft} dt \tag{D.8}$$

then

$$V(f) = \frac{1}{2} M(f + f_c) + \frac{1}{2} M(f - f_c) \tag{D.9}$$

An amplitude spectrum $|M(f)|$, band limited to the range $-f_m$ to $+f_m$, is shown in Fig.D.2a. In Fig.D.2b, the corresponding amplitude spectrum of $|V(f)|$ is shown.

D.2 Linear System Response

A *linear system* is one in which, in the frequency domain, the output amplitude at a given frequency bears a fixed ratio to the input amplitude at that frequency and the output phase at that frequency bears a fixed difference to the input phase at that frequency, irrespective of the absolute value of the input signal. Such a system can be characterized by the complex transfer function, $H(f)$ say, given by

$$H(f) = |H(f)| e^{-j\theta(2\pi f)} \tag{D.10}$$

where $|H(f)|$ represents the absolute amplitude characteristic and $\theta(2\pi f)$ the phase characteristic of $H(f)$.

Consider a linear system with complex transfer function $H(f)$, as shown in Fig. D.3, with an input signal $v_i(t)$, an output signal $v_o(t)$, and with corresponding spectral amplitude densities of $V_i(f)$ and $V_o(f)$. After transfer through the system, the spectral amplitude density of $V_i(f)$ will be changed to $V_i(f)H(f)$. Thus,

$$V_o(f) = V_i(f) H(f) \tag{D.11}$$

and

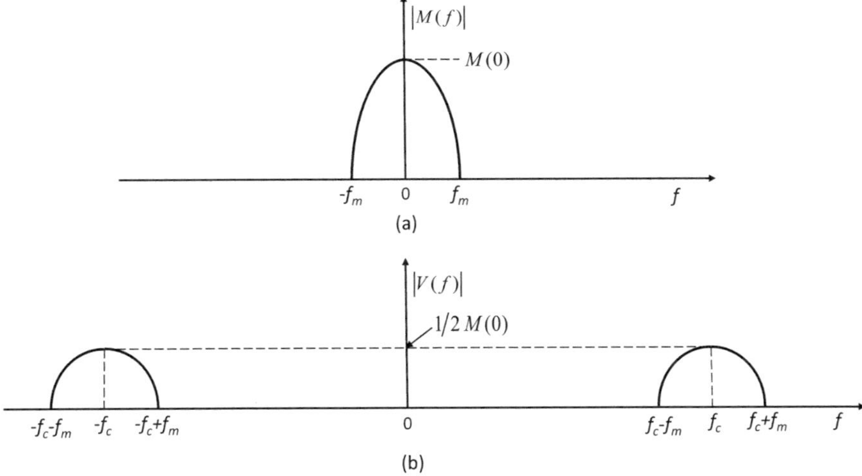

Fig. D.2 (a) The amplitude spectrum of a waveform with no special component beyond f_m. (b) The amplitude spectrum of the waveform in (a) multiplied by $\cos 2\pi f_c$

Fig. D.3 Signal transfer through a linear system

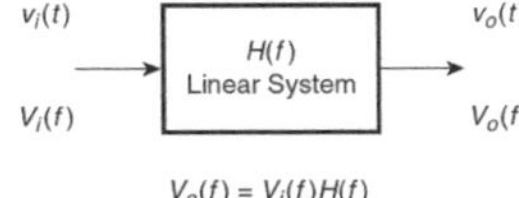

$$v_o(t) = \int_{-\infty}^{\infty} V_i(f)H(f)e^{j2\pi ft}\,df \tag{D.12}$$

An informative situation is the one where the input to a linear system is an impulse function of unit strength. For this case, as per Eq. (D.5), $V_i(f) = 1$, and

$$V_o(f) = H(f) \tag{D.13}$$

Thus, the output response of a linear system to a unit strength impulse function is the transfer function of the system.

Appendix E

QAM Cross-Constellation BER Computation

In Burr [1], a method is outlined for the determination of the *probability of symbol error P_{se}*, which in turn can lead to an estimate of the *probability of bit error P_b* for QAM systems. With this method, P_{se} is given by

$$P_{se} = \overline{n_n} Q\left(\frac{a}{\sigma}\right) \tag{E.1}$$

where $\overline{n_n}$ is the sum of the immediate neighbors of each constellation point divided by the number of constellation points, i.e., the average number of immediate neighbors,
where the spacing between nearest constellation points is $2a$,
where σ is the standard deviation of the accompanying noise,
and where $\overline{A^2}$ is the average power of the QAM signal.

$\overline{A^2}$ equals a^2 times the factor k and is thus given by

$$\overline{A^2} = k \times a^2 \tag{E.2}$$

Also given in [1] is the relationship:

$$\frac{\overline{A^2}}{\sigma^2} = 2\frac{E_s}{N_0} \tag{E.3}$$

where E_s is the energy per symbol
and N_0 is the power spectral density of the additive white Gaussian noise.
Substituting Eq. (E.2) into (E.3), we get

$$\frac{a^2}{\sigma^2} = \frac{2E_s}{kN_0} \tag{E.4}$$

Thus,

$$\frac{a}{\sigma} = \sqrt{\frac{2E_s}{kN_0}} \tag{E.5}$$

and substituting Eq. (E.5) into (E.1), we get

$$P_{se} = \overline{n_n} Q\left(\sqrt{\frac{2E_s}{kN_0}}\right) \tag{E.6}$$

If there are m bits per symbol, then the energy per symbol E_s is related to the energy per E_b via

$$E_s = mE_b \tag{E.7}$$

And hence,

$$P_{se} = \overline{n_n} Q\left(\sqrt{\frac{2mE_b}{kN_0}}\right) \tag{E.8}$$

We thus see that to determine the probability of symbol error as a function of E_b/N_0, we only need to determine the average number of immediate neighbors and the factor k, m being known.

The probability of bit error P_b is related to the probability of symbol error P_{se} via the nature of the coding. If, for example, the constellation is fully Gray-coded so that any nearest neighbor differs by only one bit, then if one symbol is generated from m bits, then one symbol error will only result in 1 bit error for every m bits received.

The method above is applicable to all QAM constellations given equal spacing in each plane between the constellation points. It is particularly helpful, however, in computing error probability in the case of *cross-constellations*. This will be demonstrated by computing the error probability of 32-QAM with the constellation as shown in Fig. E.1.

A careful study of Fig. E.1 will show that there are 16, 8, and 8 points that have 4, 3, and 2 immediate neighbors, respectively. Thus,

$$\overline{n_n} = \frac{(16 \times 4) + (8 \times 3) + (8 \times 2)}{32} = \frac{13}{4} \tag{E.9}$$

Assuming that the spacing between nearest constellations points is $2a$, then

$$\overline{A^2} = \frac{(4 \times 2a^2) + (8 \times 10a^2) + (4 \times 18a^2) + (8 \times 26a^2) + (8 \times 34a^2)}{32}$$

$$= 20a^2 \tag{E.10}$$

Thus, k equals 20, and since for 32-QAM m equals 5, P_{se} as per (E.8) is given:

$$P_{se} = \frac{13}{4} Q\left(\sqrt{\frac{E_s}{2N_0}}\right) \tag{E.11}$$

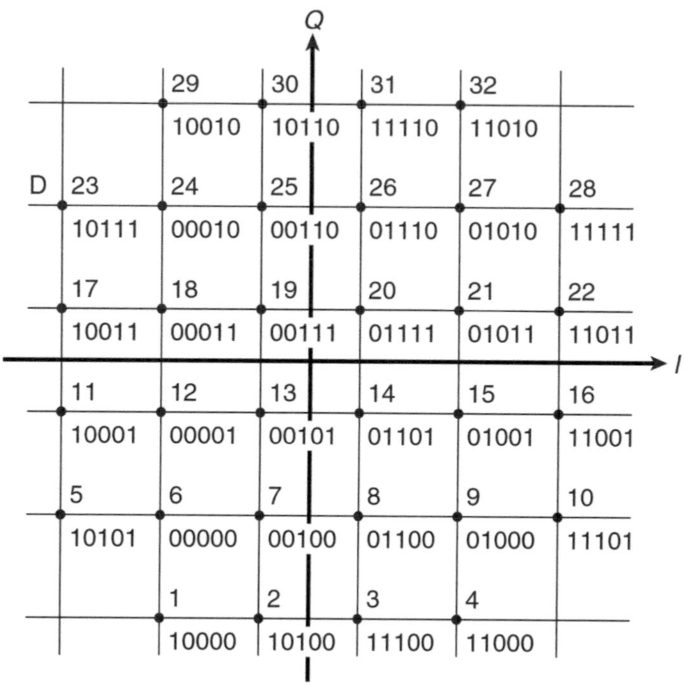

Fig. E.1 A 32-QAM constellation

We note that the constellation is not fully *Gray-coded*. Specifically, states 5, 10, 23, and 28 each differ by more than one bit for their closest neighbors. Thus, an error resulting from one of these four states being decoded as one of its closest adjacent states will result in more than one bit being in error. Since there are five bits in a symbol, the bit error rate will be greater than one-fifth of the symbol error rate. However, as most of the symbols don't differ by more than one bit from its closest neighbors, the actual bit error can be assumed to be, for practical purposes, that achieved by full Gray coding, that is, one-fifth of the symbol error rate. Thus, the probability of bit error P_b is given by

$$P_b \simeq \frac{13}{20} Q\left(\sqrt{\frac{E_s}{2N_0}}\right) \tag{E.12}$$

Reference

[1]. Burr A (2001) Modulation and coding for wireless communications. Pearson Education Limited, Harlow

Index

A

Adaptive baseband equalization fundamentals
  DFE, 159
  equalizer input pulse waveform, 155
  FIR, 155
  ISI, 156
  linear feedforward transversal equalizer,
    154
  LMS algorithms, 157, 158
  nonminimum phase fading, 159
  postcursor ISI, 157, 159
  RLS algorithms, 157, 158
  tapped delay-line equalizer, 154
  TVE, 154
  zero-forcing approach, 157, 158
Adaptive equalization, 17
  adaptive equalizers, 153
  frequency selective fading, 153
  ISI, 153
  time-domain (*see* Time-domain
    equalization)
Adaptive equalizers, 153
Adaptive modulation and coding (AMC), 141
Adaptive Multirate (AMR), 28
Air interface capacity (AIC), 198, 199
Amplitude spectral density, 222
Amplitude spectrum, 225
AMR-Narrowband (AMR-NB), 28
AMR-Wideband (AMR-WB), 28
Analog-to-digital converter (ADC), 170, 192
Angle diversity, 69, 70, 72, 73
Antenna coupler, 189, 190, 194, 196
Antenna coupling, 195–197
Antenna duplexer, 195

Antenna gain, 34
Antenna noise temperature, 106
Antennas, 198
  characteristics
    antenna gain *vs.* angular deviation, 35
    beamwidth, 35
    channel discrimination, 36
    efficiency factor, 35
    frequency, 35
    front-to-back ratio, 36
    gain, 34
    horizontally polarized, 36
    isotropic radiator, 34
    operation, 35
    pole/boresight, 35
    radiator, 35
    transmit mode, 35
    transmitting, 34
    vertically polarized, 36
    wireless transport systems, 35
    XPD, 36
  wireless communication, 34
Antenna transmit channel, 168
Application layer protocol, 20
Associated BPSK (ABPSK), 90
Asynchronous Transport Mode (ATM), 10
Atmospheric absorption, 49, 50, 52
Atmospheric effects
  absorption, 49, 50, 52
  rain attenuation, 49, 50, 52
  reflection, 49
  refraction
    advection ducts, 48
    atmospheric refraction, 44

© The Author(s), under exclusive license to Springer Nature Switzerland AG 2021   229
D. H. Morais, *5G and Beyond Wireless Transport Technologies*,
https://doi.org/10.1007/978-3-030-74080-1

Atmospheric effects (*cont.*)
   digital wireless transmission, 48
   ducting, 47
   earth's surface, 46
   electromagnetic wave, 45
   environment, 47
   evaporation, 48
   fiber optic glass cable, 48
   ground ducts, 47
   humidity, 48
   multiple refractive atmospheric paths, 48
   multiple refractive paths, 48
   radio path, 46
   radio signal, 44, 47
   radio wave, 45
   refractive index, 45
   substandard/subrefractive, 46
   superstandard/superrefractive, 46
   temperature, 48
   transmitter, 45
Atmospheric gasses, 51
Atmospheric multipath (Rayleigh) fading, 61, 62
Atmospheric multipath fading, 62
Atmospheric reflection, 49
Atmospheric refraction, 44
Automated frequency coordination (AFC) system, 12
Automatic Repeat Request (ARQ), 118

B
Backhaul, 4–15
Band and carrier aggregation (BCA), 18, 183–186
Bandpass filter (BPF), 114, 147
Baseband (BB) format, 4
Baseband data transmission
   amplitude characteristics
      Nyquist channel, impulse transmission, 77
      rectangular pulse transmission, Nyquist channel, 79
   amplitude spectral density, 76
   brickwall filter, 77
   communication systems, 75, 76
   components, 80
   digital communication systems, 77
   dispersion, 76
   eye diagram/eye pattern, 79, 80
   filter, 78
   impulse transmission, 77
   ISI, 79, 80

   non-distorted pulse, 81
   nondistorted pulse amplitudes, 78
   Nyquist criterion, 76
   PAM, 76
   receiver filters, 81
   roll-off factor α, 77
   RRC filter, 81
   time responses, 78, 79
   transfer functions
      receiver filters, 81, 82
      transmitter filter, 81, 82
Baseband processor (BBP), 189, 191, 193
Baseband unit (BBU), 6, 10
Beam bending fading, 60
Beamwidth, 35, 37, 38
Belief propagation decoding, 127
Binary discrete memoryless channel, 133
Binary erasure channel (BEC), 134
Binary phase-shift keying (BPSK), 85–87
Bit decoding true/false matrix, 130
Bit-error-rate (BER), 33, 44, 63, 66, 68, 98, 117, 211
Blind compensation techniques, 152
Blind equalization, 162
Block codes, 119–123, 132
Block interleaving
   deinterleaving, 139, 140
   interleaving depth, 140
   wireless noisy channel, 139
Brickwall filter, 77
Burst errors, 128
Butler matrix, 182
Bytes, 19

C
Carrier recovery, 16
   Costas loop method, 110
   decision directed method, 110
   differential decoder, 112
   differential encoding, 112
   digital transmission, 109
   discrete component, 110
   DSBSC systems, 110
   frequency offset, 111
   frequency synchronization, 109
   high-order QAM, 110
   incoming carrier, 110
   modern oscillators, 112
   multiply-filter-divide method, 110
   $\varepsilon°$ phase error, 16-QAM constellation, 112
   QAM receiver with decision directed carrier phase correction, 111

QAM signal, 109, 110
  received spectrum, 110
  receiver local oscillator(s), 110
  transmitted QAM signal, 109
  VCO, 111, 112
Cassegrain antenna, 37
Centralized RAN (C-RAN), 5
Central unit (CU), 3
Channel capacity, 133
Channel polarization, 133–135
Check nodes (CNs), 124, 127
Checksum, 25
Circuit switching, 19
CMA-based equalizer, 163
Co-channel dual polarization (CCDP), 13, 17,
      167–170, 179, 202, 205
Code Division Multiple Access (CDMA)
      technology, 10
Coding, 16
Coding gain, 119
Coherent detection, 84
Common Public Radio Interface (CPRI), 6
Communication path state estimator (CPSE),
      176
Communication systems, 75, 76
Connectionless protocol, 21, 22
Connection-oriented protocol, 21
Constant-modulus algorithm (CMA), 162, 163
Convolution codes, 119
Corrected bit error rate (CBER), 132
Costas loop method, 110
CRC-aided SCL (CA-SCL), 138
Cross-channel adaptive transversal equalizer
      (XC-ATE), 169
Cross-constellation, 226–228
Cross-constellation 2^{2n+1}- QAMs, 101
Cross-constellation class, 97
Cross-polarization discrimination (XPD),
      66, 67
Cross-polarization interference (XPI), 168
Cross-polarization interference cancellers
      (XPICs), 67, 168, 169, 179
Cyclic block codes, 121
Cyclic redundancy check (CRC), 25

D
Data link layer Ethernet protocol, 24–26
Data packet transmission, 29
D-Band, 13
Decision directed method, 110, 111
Decision-directed equalizer, 158
Decision feedback equalizer (DFE), 159, 161

Decode polar codes, 136–138
Decoding, 126, 127
Descrambling, 109
Diffraction, 56, 58
Digital communication systems, 77, 86
Digital heterodyne receiver, 104
Digital heterodyne transmitter, 103
Digital to analog converter (DAC), 142, 192
Digital transmission, 109
Digital wireless systems, 106, 117, 127, 164
Direct conversion, 102, 105, 109, 189, 190,
      194, 201
Dispersion, 76
Distributed RAN (D-RAN), 5
Distributed unit (DU), 3
Diversity baseband switch, 72
Diversity improvement factors, 70, 72
Double-sideband (DSB) signal, 83
Double-sideband suppressed carrier (DSBSC)
      modulation, 82–84
Downconversion, 75, 104, 109, 112, 115,
      190, 194
Dribbling errored second (Dribbling ES), 68
Dual-stack IPv4/IPv6, 24
Duct entrapment fading, 60
Ducting, 47, 48

E
Earth radius factor, 46, 58
E-Band, 13
Effective path distance, 217, 219
EffnetBHC™ scheme, 30
Electromagnetic waves, 17, 45, 179
Encapsulation, 21
Enhanced CPRI (eCPRI), 6
Enhanced Data Rate for GSM Evolution
      (EDGE), 10
Enhanced mobile broadband (eMBB), 1
Equal gain combiner (EGC), 70
Equalization, 153
Equalizer input pulse waveform, 155
Equivalent Isotropic Radiated Power (EIRP),
      12, 36
Error-control coding, 118
Ethernet
  fast Ethernet, 25
  Gigabit, 25
  MPLS, 30
  SFP, 26
  twisted-pair copper cable and RJ-45
      connections, 26
  twisted-pair technologies, 25

Ethernet devices, 25
Ethernet frame, 25, 27
Ethernet packet format, 24
Ethernet packets, 24, 25
Ethernet transported data, 20, 31
Euclidean distance *vs.* Hamming distance,
    121, 122
Euclidian distance, 121
Eye diagram/eye pattern, 79, 80

F
Fading
    atmospheric effects (*see* Atmospheric
        effects)
Fast Ethernet, 25
5G networks, 19
Fifth generation (5G) wireless transport links,
    30, 75
Finite Impulse Response (FIR) filter, 155
First generation (1G) systems, 9
Fixed analog network, 176
Fixed wireless path
    antennas (*see* Antennas)
    availability, 68
    BER, 44
    cross-polarization, 66
    digital system, 33
    diversity
        angle, 72, 73
        space, 70, 72
        techniques, 69
    error-free state, 68
    external interference, 67
    fade margin, 43
    fading/obstruction, 33, 42
    free space propagation, 40–42
    gains and losses, 42
    hydrometeors, 69
    millimeter-wave bands, 33
    non-fading environment, 44
    outage event, 68
    path reliability, 69
    radio waves, 33
    received input power, 43
    system gain, 44
    typical, 33, 34
    unacceptable performance, 68
    unavailability, 68
    worst month performance, 68
Fixed wireless systems, 132
Flat fading, 59, 60, 62, 70
Flat plane/planar array antenna, 34

Forward error correction (FEC), 167
    ARQ, 118
    BER performance, 119
    binary communications, 118
    block codes, 119–123
    categories, 119
    coding gain, 119
    convolution codes, 119
    digitally modulated signals, 118
    error detection and correction, 119
    error-control coding, 118
    LDPC codes, 118, 124–127, 133
    PCM, 123, 124
    polar codes (*see* Polar codes)
    Reed-Solomon codes, 119
    RS codes, 127, 128, 130–133
    vs. uncoded error performance, 118
Fourier transform, 222, 223
Fragmentation, 23
Frame, 25
Free space loss, 40–43
Free space optics (FSO), 1, 7, 8
Free space propagation, 40–42
Frequency channel, 167
Frequency diversity, 69, 70
Frequency division duplexed (FDD), 174
Frequency domain adaptive equalizers, 153
Frequency domain equalizers (FDEs), 153, 176
Frequency selective fading, 59–62, 70, 72
Fresnel zones, 53–56, 58
Front end, 105–107
Fronthaul, 4–11, 13
Front-to-back ratio, 36, 37

G
Generator polynomial, 25
Geography, 67
Gigabit Ethernet, 25
Global System for Mobile Communications
    (GSM), 9
Gray coded, 95, 98, 99, 227, 228
Ground reflection fading, 60, 61

H
H.264 Advanced Video Coding
    (H.264/AVC), 29
Hamming codes, 129, 130
Hamming distance, 121, 122
Hard decision decoding, 121
Header, 21
Header compression, 29, 30

Heterodyne conversion, 103, 190, 195
Heterodyne receiver, 105
Higher-order QAM systems, 93
High-order bandpass filtering, 103
High-order linear modulation, 105
High-order 2^{2n+1}-QAM, 96–99
High-order 2^{2n}-QAM, 94
    block diagram, 93
    gray coding, 95
    high spectral density, 93
    output symbols, 93
    16-QAM system, 93–96
    64-QAM system, 95
    256-QAM system, 96
    1024-QAM system, 96
    4096-QAM system, 96
High-order QAM, 110
Homodyne conversion, 102
Horizontal and vertical polarization, 67
Huygen's principle, 56

I
IF processor, 193
Image frequency, 104, 105
Impulse transmission, 77
Integrals, 210
Integrated Access and Backhaul (IAB), 14, 15
Inter symbol interference (ISI), 153
Interframe gap (IFG), 25
Interleaving depth, 140
Intermediate frequency (IF), 103, 109
Internet Engineering Task Force (IETF), 22,
    27, 28
Internet layer protocol, 22–24
Internet of Things (IoT), 2
Internet protocols, 22
Inter-symbol interference (ISI), 79, 80, 153, 176
IP datagram, 23
IPv4, 23
IPv4-based VoIP packet, 28
IPv4 datagram, 22
IPv4-to-IPv6 transition technology, 24
I/Q amplitude imbalance, 150
I/Q balance error mitigation, 150, 151
I/Q imperfections, 152
Isotropic antennas, 40
Isotropic radiator, 41

L
LAN/MAN networks, 29
Least mean square (LMS) algorithms, 157, 158,
    170, 177
Linear array, 39

Linear binary code, 120
Linear feedforward transversal equalizer, 154
Linear modulation systems, 142
    baseband signal *vs.* modulated RF carrier, 82
    BPSK, 85–87
    DSBSC modulation, 82–84
    high-order 2^{2n+1}-QAM, 96–99
    high-order 2^{2n}-QAM, 93–96
    PAPR, 100–102
    QAM, 88, 89
    QPSK, 89–93
    wireless communication systems, 82
Linear system, 224, 225
Line-of-sight (LOS), 171
Line-of-sight multiple-input multiple-output
    (LoS MIMO), 17
    arrangement, 172
    capacity, 171
    channel capacity, 177, 179
    equalization, 176, 177
    interference cancellation, 172
    LoS, 171
    millimeter-wave range, 171
    mobile network wireless transport links, 171
    NLoS, 171
    non-optimal antenna separation, 175, 176
    N receivers, 171
    N transmitters, 171
    optimal antenna separation, 173, 174
    signal analysis, 173
    signals, 171
    spatial separation, 171
    transmission, 171
    transmit and receive antennas, 171
    vertical separation, 171
Link adaptation, 17
Link capacity capability
    AIC, 198, 199
    E-band, 199
    Ethernet frame size, 200
    factor, 199
    features, 200
    single channel, 198
    transmission, 200
LO leakage, 151
Logarithmic likelihood ratios (LLRs), 123,
    127, 136
Long term evolution (LTE), 10
Low-density parity-check (LDPC) codes, 16,
    118, 119, 125
    decoding, 126, 127
    linear FEC codes, 124
    parity-check codes, 125
    PCM, 125
    QC LDPC code, 125, 126

Low-density parity-check (LDPC) codes (*cont.*)
   Tanner graph, 125
   VNs, 125
   wireless transport, 133

M
Main channel adaptive transversal equalizer
   (MC-ATE), 169
Massive machine-type communications
   (mMTC), 1
Matrix algebra, 210
Maximal ratio combiner (MRC), 72
Maximum likelihood (ML) decoding, 124
Maximum power combiner (MPC), 70, 72
Maximum transfer unit (MTU), 23
Medium Access Control (MAC), 2
Message-passing algorithms, 127
Midhaul, 4–9, 11, 13
MINI-LINK 6352, 202
Minimum dispersion combiner (MDC), 72
Minimum distance, 121, 124, 125, 128, 129
Minimum Hamming distance, 121
Mobile network transport, 201
Mobile Termination (MT), 15
Modem realization techniques, 108
   carrier recovery (*see* Carrier recovery)
   descrambler, 109
   digital modulator, 107
   scrambler, 107, 108
   timing recovery, 113, 114
Modern oscillators, 112
Modified constant-modulus algorithm
   (MCMA), 162, 163, 170, 177
Modulation methods, 75, 114
Modulo-2 addition, 120, 123, 129, 130
Monitored cold standby protection, 197
Monitored hot standby protection, 196
Multi-layer header compression, 16
Multipath fading channel model
   amplitude and group delay, 63
   channel bandwidth, 64
   channel model, 63
   digital radio channels, 64
   digital radio links, 63
   dispersion signature, 65
   flat fade power, 63
   frequency selective effects, 63
   minimum phase, 64, 65
   non-minimum phase, 64, 65
   16-QAM digital radio, 66
   quantity, 63
   radio performance, 65
   signature depth, 65
   signature width, 65
   three-ray model, 63–65
   two-ray model, 64
Multiple-input multiple-output (MIMO), 6, 13,
   170, 175, 182
Multiply-filter-divide method, 110
Multi-protocol Label Switching (MPLS), 27, 30

N
Noise figure, 106, 107
Nonbinary block codes, 127, 128
Nondistorted pulse amplitudes, 78
Non-line-of-sight (NLoS), 170
Nonminimum phase fading, 154
Non-modulation-based capacity improvement
   techniques, 167
Non-optimal antenna separation, 175, 176
Nonperiodic function, 222, 223, 225
Non-return-to-zero (NRZ), 89
Nonselective outage, 211, 212, 214
Non-square 64-QAM signal point constellation,
   102
Nyquist bandwidth, 87
Nyquist criterion, 76

O
Optimal antenna separation, 173, 174
Orbital angular momentum (OAM)
   application, 179
   characteristics, 181, 182
   demultiplexing, 182, 183
   electromagnetic waves, 179
   mobile network wireless transport, 179
   mode generation, 182, 183
   multiplexing, 17, 182, 183
   structure, 181, 182
   transmission modes, 179
Orthogonal frequency division multiplexing
   (OFDM), 10
Orthomode transducer (OMT), 38
Outage prediction, 213
Output bandpass filter, 105

P
Packet-based systems, 141
Packet data communication systems, 19
Packet Data Convergence Protocol (PDCP), 2
Packet-switched networks, 19
PAM baseband signal, 88

PAM DSBSC signals, 97
Parabolic antenna, 34, 36–39, 66
Parity-check block codes, 123, 124
Parity-check matrix (PCM) codes, 123, 124
Patch array antenna, 39
Path reliability, 44, 69
Payload compression, 30, 31
Peak-to-average power ratio (PAPR), 88,
        100, 101
    cross-constellation 2^{2n+1}- QAMs, 101
    non-square constellations, 101
    non-square 64-QAM signal point
            constellation, 102
    peak power, 102
    16-QAM constellation, 100
    rectangular constellation 2^{2n}- QAMs, 100
    unfiltered symbols, 101
Phase-locked loop (PLL), 114, 144
Phase noise (PN), 143
Phase noise suppression, 17
    BPF, 147
    heterodyne millimeter-wave systems, 147
    millimeter-wave frequencies, 146
    millimeter-wave up- and downconversion
            processes, 148
    mitigation, 147
    oscillator's phase noise, 144
    pilot-based phase noise mitigation, 147
    pilot-symbols, 146, 147
    PLL oscillators, 145
    PSD, 144, 145
    16-QAM *vs.* QPSK, 146
    16-QAM symbols, 145
    single carrier system, 145
    upconversion process, 147
Pilot-based phase noise mitigation, 147
Pilot-symbol assisted modulation (PSAM), 147
Pilot-symbols, 146, 147
Plane-electromagnetic (PE) wave, 181
Point-to-point wireless, 171
Point-to-point wireless antennas
    beamwidth, 38
    Cassegrain antenna, 37
    diameter and frequency, 38
    dipole planar array, 39
    dual-reflector type, 37
    efficiency factor, 37
    flat panel antenna, 39
    frequency, 37, 38
    linear array, 39
    OMT, 39
    parabolic antenna, 36
    patch array antenna, 39
    planar array antenna, 39
    radome, 37

RF energy, 36
rules of optics, 37
shield, 37
single- and dual-reflector parabolic
        antennas, 37
Polar coding, 16, 118
    binary discrete memoryless channel, 133
    block codes, 133
    channel capacity, 133
    channel polarization, 133–135
    decoding, 136–138
    encoding, 135
Postcursor ISI, 153, 157, 159, 161
Power amplifier linearization via predistortion,
        142, 143
Power spectral density (PSD), 144
Precursor ISI, 153, 154, 157, 159, 161
Predistortion, 142
Probability of bit error, 226, 228
Probability of symbol error, 226, 227
Protocols, 19
Pulse amplitude modulated (PAM), 76, 82, 83,
        85, 88, 97
Puncturing, 140

Q
QAM adaptive baseband equalization, 160, 161
32-QAM cross-constellation, 97, 98
32-QAM modulator logic circuit signals, 98, 99
QPSK hardware, 91
Quadrature amplitude modulation (QAM), 16,
        17, 75, 88, 89, 189, 226–228
Quadrature crosstalk, 160, 161
Quadrature error mitigation, 149, 150
Quadrature imperfection mitigation, 17
Quadrature modulation/demodulation
        imperfections mitigation
    I/Q balance error, 150, 151
    I/Q impairments, 149
    modulation degradations, 148
    quadrature error, 149, 150
    receiver, 151, 152
    residual error, 151, 152
Quadrature phase-shift keying (QPSK), 87,
        89–93
Quality of service (QoS), 23
Quasi-cyclic (QC) LDPC codes, 125, 126

R
Radio access network (RAN), 2, 16
Radio Link Control (RLC), 2
Radio over Ethernet (RoE) interface, 7
Radio unit (RU), 4, 5

Radome, 37
Rain attenuation, 49, 50, 52
Rain fading, 60, 68
Rain outage analysis, 216, 217, 219, 220
Raised cosine filter, 77, 78
Real-time Transport Protocol (RTP), 28
Receiver downconverter, 104, 105
Receiver quadrature imperfections mitigation, 151, 152
Rectangular constellation 2^{2n}- QAMs, 100
Recursive least mean square (RLS), 170
Recursive least square (RLS) algorithms, 157, 158, 177
Reed-Solomon (RS) codes, 16, 118, 119
    additive Gaussian noise, 131
    advanced hardware architecture, 131
    bit decoding true/false matrix, 130
    block codes, 132
    burst errors, 128
    digital wireless systems, 127
    encoding and decoding process, 132
    fixed wireless systems, 132
    Hamming codes, 129, 130
    information designated symbols, 128
    nonbinary block codes, 127, 128
    random symbol error block performance, 132
    wireless transport, 133
Reflection, 49, 52, 53
Reflection coefficient, 53, 61
Refractive index, 45
Remote radio head (RRH), 5, 10
Residual bit error rate (RBER), 68
Residual error, 151, 152
RF power amplifier, 142
Robust Header Compression (ROHC), 30
Roll-off factor, 77, 78
Root-raised cosine (RRC) filter, 81

S
Sampling function, 222
Scrambler, 107, 108
Second generation (2G) systems, 9
Selective outage, 212–214
Service Data Application Protocol (SDAP), 2
Severely errored second (SES), 68
Signal space/vector/constellation diagram, 85
Signal-to-noise ratio, 86, 106
Simplified constant-modulus algorithm (SCMA), 162, 163, 170, 177
Single carrier system, 145
Single-channel antenna duplexing, 195

Single-input single-output (SISO), 172
Small Form Factor Pluggable (SFP), 26
Soft decision decoding, 121–123
Space diversity, 69, 70, 72, 73
Space-time equalizers (STEs), 177, 179
Spectral efficiency, 87, 91, 94, 95, 97, 98, 117, 164
Spin angular momentum (SAM), 181
Square and filter timing recovery method, 113, 114
Successive cancellation (SC) decoding, 136–138
Suppressed carrier signal, 84
Symbol timing recovery method, 114
Systematic binary linear block encoding, 119

T
Tanner graph, 124
Tapped delay-line equalizer, 154
Terrain effects
    atmospheric effects, 52
    diffraction, 56, 58
    Fresnel zones, 53–55
    path clearance criteria, 58, 59
    radio signals, 52
    reflection, 52, 53
Terrain reflection, 52, 53
Terrestrial wireless system, 106
Thermal fade margin (TFM), 44
Third generation (3G) systems, 10
Third Generation Partnership Project (3GPP), 2
Threshold to interference (T/I) ratio, 67
Time division multiplexed (TDM), 9
Time-domain equalizer (TDE), 154
    adaptive baseband equalization fundamentals (*see* Adaptive baseband equalization fundamentals)
    frequency selective fading, 153
    initialization methods, 162, 163
    ISI, 153
    nonminimum phase fading, 154
    postcursor ISI, 153
    QAM adaptive baseband equalization, 160, 161
Timing jitter, 80, 114
Timing recovery, 16, 113, 114
Total outage, 211
Traditional data communication, 19
Traditional telephone communication, 19
Training sequence, 162
Transceiver, 213
Transceiver architecture

antenna coupling, 195–197
antennas, 189, 198
baseband, 191, 193
baseband processor interfaces, 190
BBP, 189
direct conversion, 189
direct conversion RF front end, 194
downconversion, 190
Ericsson wireless transport terminals, 191
Ethernet cable, 190
factors, 189
full outdoor mount configuration, 191
32 GHz link, 204–206
80 GHz (E-Band) link, 201–203
heterodyne conversion, 190
heterodyne RF front end, 194
IF frequencies, 190
IF processor, 191, 193
path performance, 201–207
QAM, 189
receivers, 189
specifications, 201–207
split mount configuration, 191
systems operating, 191
upconversion, 190
wireless transport transceiver, 190
wireless transport transmitters, 189
Transmission, 167
Transmission Control Protocol (TCP), 21
Transmission Control Protocol/Internet
Protocol (TCP/IP)
application layer protocol, 20
data link layer Ethernet protocol, 24–26
5G, 20
internet layer protocol, 22–24
MPLS, 27
physical data link layer, 19
protocols, 20
protocol stack, 20
TCP, 21
UPD, 21
Transmission frequency, 66
Transmission IF and RF components
baseband signal, 102
digital heterodyne receiver, 104
digital heterodyne transmitter, 103
front end, 105–107
high-order bandpass filtering, 103
output bandpass filter, 105
receiver downconverter, 104, 105
transmitter RF power amplifier, 105
transmitter upconverter, 103
Transmitter filter transfer function, 81

Transmitter *I/Q* balance error mitigation, 150
Transmitter power amplifier predistortion, 16
Transmitter RF power amplifier, 105, 142
Transmitter upconverter, 103
Transversal equalizer (TVE), 154
Trigonometric identities, 209–210

U
UDP datagram, 23
Ultra-reliable and low latency communications
(URLLC), 1
Uniform circular array (UCA), 182
Universal Mobile Telecommunications System
(UMTS), 10
Upconversion, 104, 109
User Datagram Protocol (UPD), 21
User Datagram Protocol/Internet Protocol
(UPD/IP), 19
User equipment (UE), 4

V
Variable nodes (VNs), 124, 125
V-band, 13
Video over IP, 28, 29
Video signal compression, 29
Voice over Internet Protocol (VoIP)
AMR-NB, 28
AMR-WB, 28
4G and 5G networks, 28
IP link, 27
IPv4-based VoIP packet, 28
voice payload data, 28
Voltage-controlled oscillator (VCO), 111, 112

W
W-band, 13
Wireless communication systems, 82
Wireless packet data transmission capacity, 29
Wireless transmitter designs, 142
Wireless transport
backhaul and midhaul links, 8
backhaul network, 9
components, 4–6
CU, 3
DU, 3
eCPRI, 6
eMBB, 1
evolution, 9–11
fiber, 7
fiber optic communication, 2

Wireless transport (*cont.*)
    5G networks, 1, 16–18
    FSO, 8
    IAB, 14, 15
    mMTC, 2
    mobile network, 8
    network, 2
    new radio (NR), 2
    nontraditional millimeter-wave bands, 13
    RAN, 2
    requirements, 2

traditional bands, 11–13
transmission, 9
URLLC, 2
X-Haul connections, 7
World Radiocommunications Conference 2019
        (WRC-19), 11

Z
Zero-forcing (ZF) equalizer, 157, 158